Manufacturing Strategy

The Research Agenda for the Next Decade

Manufacturing Strategy

The Research Agenda for the Next Decade

Proceedings of the Joint Industry University Conference on Manufacturing Strategy Held in Ann Arbor, Michigan on January 8–9, 1990.

edited by

John E. Ettlie
Michael C. Burstein
Avi Fiegenbaum

School of Business Administration
The University of Michigan
Ann Arbor, Michigan

Kluwer Academic Publishers
Boston/Dordrecht/London

Distributors

for North America: Kluwer Academic Publishers, 101 Philip Drive, Assinippi Park, Norwell, Massachusetts 02061 USA

for all other countries: Kluwer Academic Publishers Group, Distribution Centre, Post Office Box 322, 3300 AH Dordrecht, THE NETHERLANDS

Library of Congress Cataloging-in-Publication Data

Manufacturing strategy.

1. Manufactures—Research—Congresses. Research, Industrial—Congresses. I. Ettlie, John E. II. Burstein, Michael C. III. Fiegenbaum, Avi.
HD9720.5.M37 1990 658.5 89-26853
ISBN 0-7923-9065-2

Printed in the United States of America

Contents

Manufacturing Strategy

The Research Agenda for the Next Decade

1

INTRODUCTION: MANUFACTURING STRATEGY IN THE NEXT DECADE

John E. Ettlie, Michael C. Burstein, Avi Fiegenbaum

School of Business Administration
The University of Michigan
Ann Arbor, MI 48109-1234
(313) 936-2835

As we enter the new decade, global manufacturing is in a state of transition and restructuring. Europe 1992 promises to be a watershed year in focusing attention on inter-plant relations among manufacturers there. Further, small manufacturing in some countries might now be at risk. On the domestic scene, companies with policies of manufacturing where they sell products will continue to manufacture more products abroad, and most domestic industries are restructuring--whether they are high-tech (high R&D intensity) or not.

This *Proceedings* represents the first real attempt to capture in a snapshot the ongoing research in the area of manufacturing strategy. It comes at this most important time when manufacturing is under great pressure to change. There have been fundamental developments this decade that may eventually set aside many of the assumptions that have driven our manufacturing strategies--implied or formalized. For example, flexibility in manufacturing, when realized, changes the rule of volume dependency in most decisions made for productive capacity--whether short-range or long-range. Once this iron law of volume dependency is broken, all other issues take on a new dimension and several new horizons open. There are other developments during this period, including manufacturing taking an active role in marketing the firms' products, partnerships in manufacturing as well as product and process development, and the increasing number of managers being promoted to the executive suite who have significant manufacturing experience. All these trends promise new directions in the 1990's.

However, making appropriate choices of market commitments and technology investments to support these commitments is becoming increasingly complex and risky with the burgeoning set of options and the strength of international competition. Under these circumstances, manufacturing strategy becomes a critical link between market considerations and technological opportunities. Distinctive competences from manufacturing and from partnerships of manufacturing with other activities inside (e.g., product engineering) and outside (e.g., suppliers, customers) the firm might well determine the survival and growth of manufacturers in the decade of the 1990's and beyond. The crucial role of manufacturing strategy seems to be well recognized both in the public statements of CEO's and in the selection criteria for executives throughout manufacturing industries.

All of these developments place special responsibility upon those individuals in academia and industry alike with backgrounds and interests in manufacturing strategy. For the academicians, the training of students, particularly the industrially-experienced, in the disciplined determination of strategically-appropriate distinctive competences poses major challenges. The ability of academics to organize knowledge in this area and to find pedagogical approaches for its effective dissemination is a renewed agenda. The industry savants, on the other hand, face the difficult task of overcoming the "stockade" mentality of special interest groups within manufacturing companies to enable the development of effective organizational processes for strategic planning and associated technological change. There is the ever growing tendency to trivialize the "other" functions or disciplines in this rush to become excellent in manufacturing. We cannot allow ourselves to succumb to this temptation.

Clearly, the burden on both academics and practitioners in the "trenches" could be lightened considerably through a stronger partnership between the two groups. This *Proceedings* and the associated conference provide a vehicle for this movement to occur. The joint academic-industry papers are a sharing of insights, each academic bringing an exposure across a variety of industries, each industry author bringing a depth of experience within a particular manufacturing area. Thus, the academic-industrial combination presents the potential for that exchange of important ideas across industry lines within the manufacturing arena to which often has been credited significant industrial progress. Papers within this *Proceedings* from either academics or industrialists exclusively provide an opportunity within the larger context of the Conference for the sort of dialogue that leads to joint endeavors of the two communities. The juxtaposition of these papers should be part of the stimulus we used to plan the research agenda. Such dialogue and the associated strengthening of the partnership between these communities is one of the salient goals for this *Proceedings* and the associated Conference.

In the last decade we have witnessed the growth in the academic disciplines of strategic management and operations strategy. The field of strategic management has experienced a significant escalation in the 1980's which has been indicated by two important events: first, the publication of Porter's (1980) book on competitive strategy; second, the publication of the Strategic Management Journal. In parallel, the field of operations strategy has witnessed similar growth with the emergence of new journals and revived interest in production and operations management (POM). Therefore, it is our belief that the integration of these two academic fields can benefit each other, as well as establish new foundations for the study of manufacturing strategy.

ACKNOWLEDGEMENTS

A conference of professionals is a chore to plan and execute. We had a great deal of help along the way. Our steering committee provided ideas and inspiration as well as reviewing time and insights. Our early planning efforts were helped by advice and support from Dan Schendel and the Strategic Management Society. The Operations Management Association was a big help in getting the word out on the conference and for moral support. We had two conference coordinators: Jackie Grant and Regina Jamison. The book manuscript was prepared for the publisher with the assistance of Sheila Graham. Mary Molter was consistent in her production of the numerous mailings on the project. Zac Rolnik was a tireless editor, and we thank him for all his help. We thank our presenters and our colleagues for their patience and support, and we hope the first stake in the ground is a good one. We take sole responsibility for any errors of commission or omission that remain.

2

DEVELOPMENT OF MANUFACTURING STRATEGY: A PROVEN PROCESS

Roger G. Schroeder
Professor of Operations Management
Curtis L. Carlson School of Management
University of Minnesota
271 19th Avenue South
Minneapolis, MN 55455

T.N. Lahr
Director, Manufacturing Technical Support
3M Company
800 East Minnehaha Street
St. Paul, MN 55144

ABSTRACT

A proven process for the development of manufacturing strategy is described in this chapter. The purpose of this process is to assist managers in developing and implementing manufacturing strategies in their divisions or strategic business units. The process described here consists of ten steps including business strategy summary, examination of the external and internal business environment, and development of manufacturing mission, manufacturing objectives, and manufacturing strategies. Each of these steps is described in detail and examples are given of how this process can be used.

INTRODUCTION

Recently, American Industry has exhibited a great deal of interest in strategic planning for the manufacturing function. This has been brought about by increased competition, particularly from overseas, declining productivity growth and the need to revitalize manufacturing. As a result many large and small manufacturing firms have begun to develop manufacturing strategies including: 3M, Honeywell, FMC, IBM, General Motors, General Electric, Westinghouse, Hewlett Packard, Copeland, and others.

While this activity in manufacturing strategy in industry is quite recent, within the last few years, the literature dates back to 1969 when Wickham Skinner of the Harvard Business School published his famous article, "Manufacturing: the Missing Link in Corporate Strategy," (1). Since then, over 80 articles and books have been written as reviewed by Anderson, Cleveland and Schroeder (2). However, few of these references describe a planning process which can be used for strategic planning in manufacturing, the subject of this paper.

In order to conduct strategic planning for manufacturing it is necessary to have a planning process. The process should describe the steps to be used, who should be involved, what information is needed, and what outputs should result. The process should be simple enough so that it can be easily followed, and it should also be integrated with business planning and corporate planning. The purpose of this paper is to present such a process that the authors have developed and used in practice.

For the last three years we have been using the process described here in a variety of settings in large and small companies, and in manufacturing and service industries. Managers have found the process to be useful for developing manufacturing strategy in actual practice in over thirty different business units that we have assisted. Some of the benefits of this process are as follows.

1. Helps the business compete successfully
2. Guides tactical decision making in manufacturing
3. Helps to cope with a changing environment

4. Provides a long-run view of manufacturing
5. Enhances communications with other functions
6. Puts manufacturing in a proactive mode.

The first, and most important, reason for manufacturing strategic planning is to help the business compete. Strategic planning is externally oriented and can help manufacturing gain a competitive position or maintain its current lead. Manufacturing strategic planning is also intended to guide the day-to-day tactical decision making which goes on in manufacturing. Tactical decisions can be put into a strategic context and an overall policy framework. As such strategic planning helps set priorities among daily activities by establishing long-range objectives and strategies. This should result in a better utilization of manufacturing resources.

Whenever the environment of the business is changing, strategic planning becomes all the more important. In these cases staying with the present strategy and technology can be disastrous as the external world changes. Strategic planning in manufacturing is intended to identify the external environment and adapt to it in a strategic sense.

Many manufacturing people want to take a long-run view toward manufacturing and their jobs, but they are bogged down in day-to-day details. Manufacturing strategic planning provides the opportunity to get out of the current rut and to think about what the future holds and what manufacturing must do to help the business gain a competitive advantage.

By using a broad planning process which involves other functions in the company, manufacturing can significantly improve its communications with other functions. The planning process will help others understand where manufacturing is headed and why certain objectives and strategies are being pursued. In surverys of managers who have used our process this is often the most valuable part of the planning effort.

Finally, strategic planning for manufacturing casts manufacturing in a proactive role. Manufacturing is often caught on the defensive by reacting to other peoples plans. By doing strategic planning, manufacturing can get out in front and determine its own destiny. Manufacturing can also explain more clearly why certain things cannot be done, because they do not fit with the manufacturing strategy.

Before beginning strategic planning for manufacturing it is important to have a definition of manufacturing strategy in mind. The following definition is offered:

> **Manufacturing strategy provides a vision for the manufacturing organization based on the business strategy. It consists of objectives, strategies and programs which help the business gain, or maintain, a competitive advantage.**

Two things in this definition should be highlighted. A manufacturing strategy is more than just a plan. It should provide a vision for where the organization is headed. Manufacturing management must provide the leadership to develop the vision. Second, the manufacturing strategy should contain long-range objectives, as well as, strategies and programs for manufacturing. Thus a very broad definition of manufacturing strategy is proposed here.

The manufacturing strategy must, of course, be linked closely with the business strategy and other functional strategies. This is accomplished by explaining the business strategy as the first part of the manufacturing strategic planning process and by involving various functional people from outside manufacturing in the process.

STRATEGIC PLANNING PROCESS

Our strategic planning process for manufacturing consists of the following ten steps. Figure 1 shows the interrelationship of these steps.

1. Business Strategy Summary

2. Manufacturing Mission
3. Manufacturing Objectives
4. External Analysis
5. Internal Analysis
6. Competitive Position
7. Ideal Manufacturing
8. Critical Issues
9. Manufacturing Strategies
10. Manufacturing Programs

This process is best carried through by a planning team of about ten to twenty people. This team should consist of the manufacturing head (V.P., Director, etc.), manufacturing staff (materials control, process, etc), plant managers, engineering, laboratory, quality, marketing, accounting, and human resources. The General Manager of the division should be involved in, at least part, of the planning effort. The group assembled should be able to provide an outside, as well as, an inside view of manufacturing. In addition, a facilitator, familiar with the planning process and methodology, is needed to keep the meeting on track and to prevent a loss of consistency and direction.

Generally, a two-day retreat is scheduled to carry out this planning effort. The first day is devoted to steps 1 through 7 and the second day to steps 8 through 10. Much of the planning should be done in small group meetings with about ten people or less. This allows time for interchange of ideas and discussion. Frequently, the nominal group technique is used as a brainstorming method in order to get maximum participation (3).

Before the retreat, some advance preparation is needed. This includes preparing a draft of the manufacturing mission statement and the manufacturing objectives. Since these can usually be derived from the business plan, they can be prepared in advance, presented, and then discussed at the meeting. The other items in the planning process (steps 4-10) are best developed at the retreat itself because they require considerable input and thought. The business plan, manufacturing mission and manufacturing objectives can be sent out to the participants in advance of the retreat, if desired.

At the planning retreat, the first thing that should be done, after explaining the agenda, is to describe the business strategy and its implications for manufacturing. This should be done by the General Manager, if possible. This will not only involve the General Manager in the planning process, but provides a chance for those present to discuss manufacturing issues with the general manager of the division. It is especially important to get the general managers views concerning the **implications of the business strategy for manufacturing.**

Manufacturing Mission

The manufacturing mission states a purpose for manufacturing and a priority among objectives. It is derived from the business strategy and the general situation facing manufacturing. Usually, the mission statement is quite broad and stated in general terms. It normally does not contain any numbers or specific objectives.

An example of a mission statement is as follows:

> "The mission of manufacturing is to achieve the lowest unit manufacturing cost relative to competition. This will be done without sacrificing our high standards for quality or service to the customer. We will pursue the latest process technology and a people-based competitive advantage in order to be the low-cost producer."

Note, how this mission statement provides a clear sense of priority - low cost is being stressed. Also, the low cost objective is stated relative to the competition in order to put things into

Figure 1
Diagram of the Strategic Planning Process

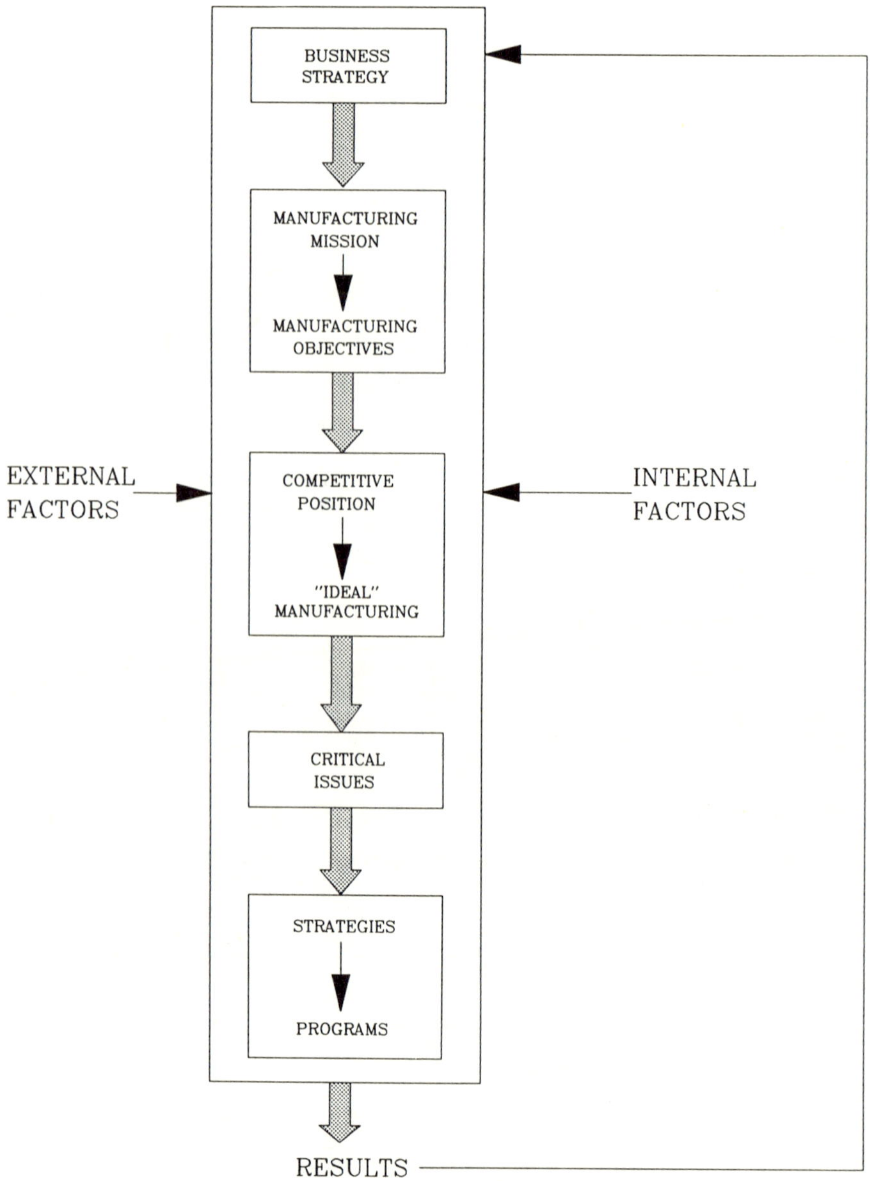

perspective. A very general indication is also given of how the objective will be achieved - through process technology and people.

Cost is, of course, not the only thing which could be stressed by manufacturing in a mission statement. Other possibilities include: quality, service, fast new product introduction, short cycle time or combinations of these. What is being stressed will generally follow directly from the business strategy.

While missions statements are general in nature, they can still convey meaning and intent. Nevertheless, too much detail should be avoided. Further refinement of the mission will take place through discussion of objectives, strategies, and competitive position. About thirty minutes to one hour, at the most, should be enough time to present and discuss the manufacturing mission and the objectives at the retreat. Total agreement should not be sought in the beginning. Some issues can best be deferred until later. Often, the mission statement will become clearer as the planning process proceeds.

Manufacturing Objectives

Manufacturing objectives consist of the performance measures that manufacturing must achieve. They should be stated in quantitative terms. Objectives could include:

1. Manufacturing Cost (as a percent of sales and/or unit costs)
2. Quality (Conformance) may be expressed as a cost of quality, or in terms of customer satisfaction
3. Inventory Turnover (or Months of Inventory)
4. Customer Service (expressed in various ways such as: % backorder, average service time, promise date compliance, etc.)
5. Cycle Time (the time from receipt of raw materials until the product is in the customers hands)
6. Time to introduce new products (from concept to initial production)
7. Time to change capacity by a stated amount from plan. (e.g. how long does it take to add 20% to planned capacity)

These objectives should ordinarily be stated on an annual basis for at least five years. The objectives should be consistent with the financial results stated in the business plan. This will usually require the objectives to be put into a financial planning model to check that they are indeed consistent with return-on-capital employed, profitability and other numbers stated for the business.

The objectives should also incorporate any other corporate goals, which are important. These corporate goals will be imbedded within the manufacturing objectives. Examples include special cost reduction targets, quality objectives, emission reductions, etc.

Objectives should be evaluated relative to the competitive situation and the requirements of the business plan. Objectives which seem rather easy to achieve may be adequate when gauged against these criteria. Other very challenging manufacturing objectives may be insufficient to do more than stay even with competition. Manufacturing should be judged by external criteria, not internal criteria. Therefore, a certain percentage improvement each year has no meaning in its own right. Meaning is only derived relative to the business plan or relative to competition.

External Analysis

External analysis is intended to help project the external environment of manufacturing into the future. The external environment is usually defined as outside of the company itself, and includes several factors such as: competition, technology, customers, economic conditions, social climate, and government regulation. These factors are considered to be largely outside the control of the company. The environment is projected for 5 or 10 years into the future, depending on the length of the planning horizon desired.

The external analysis is extremely important because it determines the parameters that manufacturing must operate within. Competition is, of course, one of the key environmental factors, since a manufacturing strategy is being developed in order to gain competitive advantage. Also, the changing needs and demands of the customer must be considered, along with emerging technologies, economic projections, social changes and changing government requirements.

It is advantageous to do environmental analysis in small group meetings. Because of the subjective and uncertain nature of the environment, maximum input is needed from various individuals with different viewpoints. A form of brainstorming, the nominal group technique, has been found useful for this purpose and is described in reference (3).

As a result of the environmental analysis a list of key environmental factors will be produced. Typical factors are as follows:

1. Environmental impact of using hazardous chemicals will force manufacturing to reduce emissions.
2. Excess Capacity in the industry will tend to create price competition and pressure on manufacturing costs.
3. New product life cycles are becoming shorter.
4. Manufacturing must rationalize its capacity on a global basis. Off-shore manufacturing is cheaper.
5. Employees want more out of their jobs. They not only want a paycheck, but a viable career path and more control over decisions which affect them.
6. Customers are demanding JIT deliveries and more customized products.

As part of the environmental analysis, it is important to not only list the factors, but to gain some understanding of how these factors will affect manufacturing.

Internal Analysis

An internal analysis is done to determine the strengths and weaknesses of manufacturing, and those of the company which affect manufacturing. This requires a careful consideration of how resources are deployed and utilized. Also, strengths and weaknesses need to be defined relative to manufacturing objectives and the external environment. What is a strength for one manufacturing organization may be a weakness for another with a different set of conditions.

The strengths and weaknesses are stated for current conditions, at the present time. Generally speaking, management is usually quite aware of its strengths and weaknesses, although there may be differences of opinion. Input from outside manufacturing is especially helpful in getting a variety of points of view. Thus a small group brainstorming technique with wide representation is also recommended for doing internal analysis.

A typical list of <u>strengths</u> is as follows:

1. A global manufacturing network has been constructed which is resistent to currency fluctuations.
2. The company has a strong corporate technical base which the various divisions can utilize.
3. R&D, Manufacturing and Marketing work well together as a team to introduce new products and to solve problems as they occur.
4. The division has a strong process development function. A good pilot plant exists.

Any particular strength could also be a weakness depending on your objectives and mission. For example, item 2 above is a strength if you can utilize the corporate technical base to develop new technologies for your division. On the other hand, if your division does not need the corporate technical base, then item 2 may be a weakness since it represents higher overhead. Strengths and weaknesses should be decided relative to the environment and the manufacturing objectives.

Some typical internal weaknesses which a division may face in its manufacturing function are:

1. Excess production capacity leading to high costs.
2. A long time is required to introduce new products.
3. Manufacturing cycle time is too long.
4. We have too many conflicting priorities. No clear sense of direction exists.

Strategy should draw on strengths and seek to minimize or eliminate weaknesses. An internal analysis is, therefore, helpful in identifying possible strategies which may be needed. Before identifying these specific strategies, however, it is helpful to consider the competitive position of manufacturing and the ideal manufacturing system.

Competitive Position

The distinctive competence of manufacturing is what manufacturing does best relative to the competition. This can be stated for both present and future time periods.

One way which has been found useful to identify manufacturing distinctive competence, and incompetence, is to conduct a competitive position analysis. This analysis determines how manufacturing compares to each of the major competitors, both current and future, according to various dimensions of manufacturing. While the appropriate dimensions will vary from one business to the next, an example is as follows:

	Competitors			Our
	A	B	C	Company
Unit Cost	10	7	4	9
Quality	6	3	7	10
Service	6	10	2	9
Capacity	10	5	3	3
Process Technology	7	8	8	10
Market Share (%)	28	21	16	35

Competitors are ranked on a 1-10 scale with 10 being given to the best competitor on that dimension. Other competitors are then ranked relative to the best one. Market share percents are also included for perspective. The grid is usually completed as a group process, but could be done before hand, if desired. Inspection of the scores quickly points out areas of competitive advantage and disadvantage.

As noted earlier, other dimensions can be used. For example, Hayes, Wheelwright and Clark (4) suggest cost, quality, dependability, flexibility and innovation.

In evaluating the competitive position it is not only necessary to consider the scores on each dimension, but also the relative importance of the dimension to competitive success. For example, a high level of innovation in a commodity market may not be as valuable as low unit cost. Development of the relative value of the various dimensions can be done as part of the group retreat, using a voting technique. This is often helpful in crystallizing the group's thinking on competition and competitive priorities.

If you have one credible competitor, then it unlikely that you will achieve a score of ten on every dimension. Each competitor will develop its own means of differentiation and associated distinctive competence for manufacturing. Also, since tradeoffs are involved in the design of any manufacturing system, choices must be made about which manufacturing performance objectives are the most important. Therefore, it is likely that a credible competitor will develop at least one dimension of manufacturing performance which is better than yours. However, a weak competitor who makes poor choices may end up at a disadvantage on every dimension.

Competitive position analysis will give manufacturing a sense of what it currently does well and what it must do especially well in the future to be competitive. This helps establish priorities, since distinctive competence is something which must be achieved for survival. Manufacturing strategy hinges very heavily on this concept. It also allows the organization to identify, and focus on, areas of competitive disadvantage. These must be corrected whenever they are important to the manufacturing mission.

Ideal Manufacturing

Ideal Manufacturing is a concept which is used to establish the best or ideal way to do manufacturing at the end of the planning horizon (5 or 10) years. Current plants, equipment, human resources, and processes are ignored for the purposes of doing an ideal analysis. The question is what is the best way to supply the future products, if we were to start over. Ideal manufacturing analysis is done in order to stretch manufacturing thinking and to push beyond the current way of doing things.

As part of this step it is helpful to have a projection of the products and volumes which will be required at the end of the planning horizon. The head of R&D might be asked to present current product development plans plus make a forecast of future products and volumes. It is difficult, if not impossible, to design the ideal manufacturing system without projecting what products will be required.

The following questions should be considered in deciding on ideal manufacturing:

1. What to make and what to buy?
2. What process technology should be used?
3. What capacity is needed, how large should the plants be, and where should they be located?
4. What types of skills are needed and what kind of people environment should we have?
5. What kinds of control systems should be used for production control, quality control, and cost control?
6. What type of organization structure is best?
7. How should we deal with suppliers and customers?
8. What kind of information system should we have?

Ideal manufacturing is a useful technique in developing manufacturing strategy, because it allows managers to consider the best way of doing things in the long-run without being hampered by present constraints. A powerful reason to consider ideal manufacturing is that your competitors or potential competitors will if you don't. Also, it has been found that working toward the ideal will generally produce more positive ideas than incrementally trying to improve present manufacturing. It is more likely that a breakthrough will be developed by considering the ideal. In general, a knowledgeable group can define ideal by group discussion in 1 to 1 1/2 hours.

Critical Issues

As indicated earlier, the foregoing steps 1-7 can usually be completed on the first day. At this point, a small group, possible only the facilitator, can identify and state the critical issues. Critical issues are those gaps, or shortfalls, in performance, capability, organization, etc. which must be addressed to enable the manufacturing organization to fulfill its mission and meet its objectives. Major sources for critical issues are:

External threats
Internal weaknesses
Competitive disadvantage
Deviation from ideal

Assuming a good job has been done in the first day sessions, all major critical issues should be apparent in one or more of the four categories.

Critical issues should be limited to four to six statements of major gaps or shortfalls. They should be described in enough detail so that they can be understood by all and are, by their nature, stated in negative terms, for example:

> We don't have enough capacity to meet forecast volumes, assuming operations at present rates and yields.

The preferred method is to develop these issues between the first and second days so that they can be presented to the group, discussed and either agreed to or changed, before strategy formulation begins on the second day. Critical issues, like most of the steps in this process, are a powerful priority-setting device.

Manufacturing Strategies

Manufacturing strategies determine <u>how</u> manufacturing is going to reach its objectives within the future environment. This involves correcting weaknesses and building upon strengths. In the process of pursuing these strategies a distinctive competence will be developed. Manufacturing strategy is thus derived from all of the above steps and builds upon them.

The manufacturing strategies needed will become apparent as you go through the process. Each of the foregoing steps can lead to strategies:

1. Manufacturing strategies are needed to respond to business strategies or corporate guidelines.
2. Comparison of present levels of manufacturing performance with future requirements indicates a need for a strategy, e.g. manufacturing unit cost must be reduced by 25% over the next 5 years.
3. Strategies are developed to correct present weaknesses or to exploit strengths.
4. Strategies are developed to cope with anticipated environmental changes, e.g. more stringent pollution requirements.
5. A particular distinctive competence is needed which is currently not available.
6. Strategies are needed to reach the ideal manufacturing state.

As indicated above, critical issue statements represent the most fertile areas for strategy development. Strategies to address each of these should be developed.

As a check on the development of manufacturing strategies the following list is useful to be sure that nothing is overlooked. Strategies can be developed to deal with any or all of the following topics.

1. Make or Buy
2. Process Technology
3. Plant Capacity and Location
4. Production & Inventory Control
5. Quality Assurance
6. New Product Introduction
7. Manufacturing Organization
8. Management of People
9. Suppliers
10. Just-in-time
11. Information Systems

It is generally necessary to distinguish between objectives and strategies. In this paper objectives refer to performance dimensions of manufacturing such as cost, quality, service, flexibility, etc., while strategy refers to <u>how</u> those objectives will be achieved. Therefore, "reducing unit cost" is an objective for manufacturing and not a strategy. Any number of strategies might be used to reduce unit costs.

As a general rule manufacturing should not have more than about six strategies. Even two or three strategies is sufficient. Beyond six, there are simply too many strategies for the division to give proper emphasis and support. After all, strategies are supposed to give a sense of priorities to manufacturing. This cannot be done if there are too many strategies.

Some examples of manufacturing strategies are as follows:

1. Support fast new product introduction.
2. Develop sourcing arrangements for the new "electronics hardware".
3. Accelerate human resource development.
4. Improve process technology
5. Rationalize product assignment to plants on a global basis.
6. Implement Just-in-time Operations

These strategies will help manufacturing reach the previously stated long-run manufacturing objectives and will resolve the critical issues.

Programs

Last, but not least, programs are developed to implement strategies. Programs are generally formulated for the next one or two years. They are specific actions that will be taken to begin implementing the strategy. Later, other programs may also be developed to continue implementation; three, four, or five years out. Programs are assigned to specific individuals for accomplishment and resources are identified for programs along with a time schedule. Each strategy may have one or more programs associated with it.

Programs are the link between strategy and tactics. Programs can be budgeted and tracked for accomplishment. Strategies are more general and can only be realized through accomplishment of specific programs. Programs are tactical in nature and can be changed or adjusted to achieve the strategies.

In a strategic planning workshop it is only necessary to identify the programs which are needed and to specify responsibility, resource requirements and schedules. Follow-up action will be needed after the workshop to more carefully specify programs and to track their implementation.

If possible, programs should not be specified to study something. In some cases, of course, more study is needed, but this can also be a way to avoid a decision. Action programs should be identified whenever possible.

Programs need not be new. It is best to identify programs which are needed to achieve a given strategy without reference to whether these programs are existing or not. In some cases, existing programs will suffice. In other cases a change in emphasis of existing programs is needed, or a new program must be started.

Identifying programs is a very concrete step and what some consider the result of the strategic planning workshop. Even if no new programs are identified, however, the workshop can be a success in improving communications among the participants, in better understanding the priorities among programs, and in linking programs to manufacturing and business strategy.

IMPLEMENTATION

Implementation of strategy to achieve effective manufacturing is the desired result of manufacturing strategic planning. Most strategic planning identifies changes which are needed to adapt to the future environment and to meet manufacturing objectives. If these changes are not effectively implemented, the strategy will fail.

Manufacturing people are generally good at implementation. Nevertheless, follow-up actions are necessary, not only to implement programs, but to make changes in manufacturing strategy as needed. The strategy, itself should not change every year, or there will be little chance of success. On the other hand, the tactics or programs should be continually adjusted as circumstances dictate.

To implement manufacturing strategy regular evaluation and review is needed. A quarterly review session should be scheduled for program review. Programs should be evaluated for progress and possible corrective action. An annual session should also be scheduled to review the entire manufacturing strategy. At this session, any major changes in the environment (external or internal) should be reviewed, along with new objectives or mandates, which might influence manufacturing strategies or their implementation.

These reviews should be a regular part of management of the manufacturing department and not something separate which is tacked on to existing reviews. A new dimension can be added to existing reviews, which is to ask how progress or changes will affect our manufacturing strategies. Or better still, the new reviews can be substituted for some of the existing short-term fire fighting meetings.

The ultimate objective is to reach "strategic management" of manufacturing. This occurs when every-day activities are continually evaluated on the basis of their effect on strategy. When management in manufacturing takes a strategic view toward manufacturing, as they run the business, then the objective will be achieved. Strategic Planning is only a first step toward the larger objective of strategic management of manufacturing on an ongoing basis to achieve a true competitive advantage.

CONCLUSIONS

Manufacturing strategy can be a powerful force in a business if it is properly developed and implemented. This chapter has described a proven process for achieving an effective manufacturing strategy in a division or business unit. The process consists of ten steps including assessment of the internal and external environment, and formulation of manufacturing mission, objectives, strategies and programs.

Many companies do not have well developed manufacturing strategies. Therefore, the challenge is substantial to build effective manufacturing strategies in order to help firms gain competitive advantage. It is hoped that the process described here will help managers develop and implement manufacturing strategies in their businesses.

Further studies are needed to propose and evaluate processes for the development and implementation of manufacturing strategies. This paper is one of the few which describes a process in detail. It would be helpful if others were to propose and elaborate on processes which can be used by firms. This should help accelerate the effective use of manufacturing strategy in practice.

REFERENCES

1. Skinner, Wickham, "Manufacturing - Missing Link in Corporate Strategy, Harvard Business Review, May-June, 1969.

2. Anderson, John, Cleveland, Gary, and Schroeder, Roger, Operations Strategy: A Literature Review", Journal of Operations Management, to appear.

3. Delbecq, A. L. and Van de Ven, A. H., " A Group Process Model for Problem Identification and Program Planning," Journal of Applied Behavorial Science, 1971, 7, 466-492.

4. Hayes, Robert H, Wheelwright, Steven C., and Clark, Kim B., Dynamic Manufacturing, Free Press, 1988.

3

EUROPE 1992: IMPLICATIONS FOR MANUFACTURING STRATEGY

Emma Jane Riddle and Darrell Parker

School of Business Administration
Winthrop College
Rock Hill, SC 29733
(803) 323-2186

ABSTRACT

The Single Market program of the European Community will require major changes in corporate and manufacturing strategies. Global competitive experience suggests that global or EC-wide corporate strategies will be the most successful. With reference to the Hayes-Wheelwright analysis of stages in manufacturing's strategic role, a Stage 3 or Stage 4 manufacturing strategy is required to support a global corporate strategy. This paper describes the necessary changes in manufacturing strategy in terms of the classical manufacturing strategy decisions.

INTRODUCTION

Europe 1992, the Single Market program of the European Community (EC), is already making Western Europe a more competitive market. The stated objective of the EC is to integrate the economies of its twelve member countries. Competition will increase further as the process of removing customs duties, eliminating border controls, and adopting reciprocal product standards continues.

The Single Market program creates a need for major revisions in the manufacturing strategies of many firms. This paper describes the necessary changes in terms of the classical manufacturing strategy decisions. The Hayes-Wheelwright analysis of stages in manufacturing's strategic role is used as a framework for discussing the probable impact of the changing European environment on manufacturing strategy.

IMPLICATIONS OF THE SINGLE MARKET FOR MANUFACTURING STRATEGY

Strategic Issues

This section identifies manufacturing strategy issues related to three aspects of the Single Market: harmonization of product standards, easing of border controls, and elimination of customs duties. The cumulative effect of changes in these three areas will be a major increase in competition within the EC.

The harmonization of product standards will be based on a series of reciprocal agreements to be negotiated between now and the end of 1992 [14]. Once a standard or a range of acceptable standards for a category of products has been agreed to, any product approved for sale in one EC member country can be sold in any member country. Harmonization of standards will reduce the number of product variants required to operate throughout the EC and will permit some manufacturers to market their products in countries that have been closed to them in the past. Harmonization of product standards will impact manufacturing strategy in several areas: capacity planning, facility location and specialization, technological change, and sourcing policies.

Television sets provide a good illustration of the need for harmonizing product standards [13]. Europe now has two standards for television reception and four standards for audio signal interference. To meet these varied standards, Philips of the Netherlands turns out seven different types of TV sets for the European market. Models destined for different countries must be manufactured with different semiconductors, tuners, and even plugs. Philips estimates its costs for these product variants at twenty million dollars per year. The primary cost factors are extra engineering staff, varied component inventories, and weekly modifications to assembly lines. The Philips case demonstrates the potential for substantial product redesign and cost savings resulting from the harmonization of product standards.

The easing of border controls within the EC will eliminate lengthy and costly border delays. Two other factors -- the partial deregulation of transportation under the 1992 program and the completion of the English Channel tunnel in 1993 -- will also increase mobility across national boundaries [14]. Increased mobility will expand the number of potential sources for manufacturers, increase supplier reliability (by reducing border delays), and reduce the number of supply points required to serve the EC. Development of new channels of distribution seems likely. Manufacturing strategy will be impacted in the areas of facility planning, supplier relationships, and vertical integration.

Finally, the easing of customs duties will reduce the cost of supplying several EC countries from one supply point. Companies which can compete effectively throughout the EC will reap an immediate benefit. However, some firms may face major changes in their corporate and manufacturing strategies. For instance, a firm which has previously competed on flexibility and service may now face intense price competition, which will dictate a more cost-conscious production system. If the required cost reductions are dramatic, virtually every aspect of manufacturing strategy will be affected.

Thus, at first glance, we would expect the Single Market to have several effects. First, the Single Market will present numerous opportunities to realize economies of scale. Second, the Single Market will favor competitors who successfully implement EC-wide strategies. Third, the Single Market will affect all areas of manufacturing strategy. These expectations are supported by experience in global competition, as discussed in the next section.

Lessons from Global Competition

Manufacturing strategy is a functional strategy and must support and enhance the corporate strategy [7], [12]. Thus, in considering manufacturing strategy, it is relevant to ask which corporate strategies are likely to succeed in the Single Market. By permitting the free flow of goods, the Single Market program will encourage the development of corporate strategies that are EC-wide in scope. The impact of EC-wide strategies in western Europe is analogous to the effect of global strategies in the world market.

Multidomestic versus global strategies. In the world market, competition is multidomestic in some industry segments and global in other segments [6]. The choice of either a global or multidomestic strategy constrains strategic choices in the various functional areas: marketing, finance, manufacturing, and human resources.

In industry segments where competition is multidomestic, each company develops separate strategies for each of its foreign markets [8]. The multinational headquarters controls financial and marketing policies. Research and development may be centralized, along with some component production. However, the company's foreign subsidiaries are essentially autonomous units. Each subsidiary is treated as a profit center and makes independent decisions about strategy and operations within broad corporate guidelines. Procter and Gamble, Alcoa, and General Foods are examples of successful multidomestic companies [8].

In other industry segments, competition is based on global strategies. The global competitor must compete effectively on a world-wide basis, while being sufficiently responsive to local market needs [8]. The global competitor makes strategy decisions on a centralized basis, considering the positions of major competitors throughout the world. Decisions about marketing strategy and price are based on global competitive considerations, rather than on the competitive environment in a subsidiary's home market. The global competitor may temporarily forego current earnings in order to seize market share in a competitor's home country or to develop a global distribution system that competitors cannot match [6], [8]. Manufacturing and distribution systems are configured on either a world-wide or a regional basis, rather than on a national basis [3], [4] The success of a global strategy depends on the competitor's ability to sustain the required improvements in technology, manufacturing structure, and distribution [6], [8]. Examples of successful global competitors include DuPont, IBM, Sony, Honda, Matsushita, and Michelin [8].

A global competitor can use its full range of marketing, production, and financial resources in each national market. Therefore, a multidomestic company which is confronted with a global competitor is likely to be only partially effective unless the multidomestic company also develops a global strategy. For instance, during the early 1970's, Michelin used cash flows from its strong European operations to finance expansion into the American market [6]. Goodyear fought back by moving into the European market, reducing Michelin's financial and marketing position in its home base. Although Michelin remained a significant factor in the American market, Goodyear forced Michelin to expend additional resources defending its position in Europe and to slow down its expansion in the United States.

Cvar identified seven "triggers" which support the development of global strategies [3]. At least four of these triggers -- rationalized distribution channels, changes in government policy, product redefinition, and consolidated sources of supply -- are likely to be present in the Single Market. Since any one of these triggers may be sufficient to initiate global competition, the presence of four such triggers in the EC will provide a powerful impetus for the development of EC-wide or global strategies.

Competitive options in the Single Market. Competitive options will be discussed in terms of Porter's typology of international competitive strategies [9]. Porter identified five generic international strategies: global cost leadership, global differentiation, global segmentation, national responsiveness, and protected markets.

Two of these strategies -- protected markets and national responsiveness -- are inherently domestic or multidomestic in nature. Strategies based on protected markets are likely to be obsolete in the Single Market. National responsiveness strategies, which depend on meeting specific needs of a local market, will succeed only where there is genuine differentiation based on national tastes or needs. In some industries, such as household appliances, harmonization of product standards will remove the basis for national responsiveness strategies. In industries which depend on national tastes, such as food and apparel, national responsiveness strategies may be effective for a period of time. However, as the European market becomes more homogeneous, the effectiveness of these strategies may diminish. Therefore, companies should explore opportunities to convert national food or apparel items into products with appeal outside their country of origin. This approach can be used to convert a national responsiveness strategy to an EC-wide segmentation strategy.

Porter's remaining generic strategies can be applied either globally or throughout the EC. A competitor who pursues a global or EC-wide cost leadership strategy can realize economies of scale in research and development, procurement, and production. For differentiation strategies, competitive participation at the EC or global level can reduce the cost of differentiation [9]. If global experience is any indication, EC-wide cost leadership or differentiation strategies will require substantial financial resources and the ability to preempt competition [3], [6], [8]. Smaller companies may be more successful with EC-wide segmentation strategies. L. M. Ericsson of Sweden, a supplier of telephone hardware and software to small phone companies, has demonstrated that smaller companies can successfully practice large-scale segmentation strategies, even on a global level [8]. Companies who already pursue segmentation strategies may benefit from the Single Market, since their market segments will be larger.

Implications for manufacturing strategy. We have seen that the Single Market favors the development of EC-wide strategies. The experience of global competitors suggests that EC-wide strategies will require more coordinated production systems than domestic or multidomestic strategies. Global competitors have reduced costs by coordinating procurement across national boundaries, by sourcing a plant in one country with components manufactured in another, by handling aggregate production planning on a global level, and by making capacity and facility planning decisions on a global or regional basis [3], [4], [11]. Global companies have also capitalized on technical synergies by transferring new process technology and manufacturing engineering expertise from one country to another. Similar types of coordination will be needed to obtain the maximum benefit from EC-wide strategies. A final characteristic of successful global competitors is that their organizations are centralized enough to maintain effective control and flexible enough to permit adaptation to local conditions and effective transfer of technology from one national subsidiary to another [1], [4].

Cvar identified five characteristics of successful global competitors [3]. Several of these characteristics are directly related to manufacturing strategy. Successful global competitors explicitly coordinate important value chain activities, such as production and distribution. These competitors also invest heavily in research and development, spending an average of forty per cent more than industry competitors. Successful global competitors measure themselves by global, rather than national, standards of performance in such key areas as cost and quality. For instance, in recent years the American automotive industry has moved toward global standards of quality. It is reasonable to expect that successful EC-wide competitors will have characteristics similar to those of successful global competitors.

Stages in Manufacturing's Strategic Role

Hayes and Wheelwright identified four stages in the development of manufacturing's strategic role [7]. Stage 1 or "internally neutral" companies focus on minimizing the negative potential of manufacturing, primarily through the use of planning and control systems. Stage 2 or "externally neutral" companies seek to achieve manufacturing parity with competitors and regard capital investment as the primary means for doing so. Stage 3 or "internally supportive" companies relate manufacturing decisions to the business strategy and systematically address longer-term manufacturing trends and new process technologies. Their view of manufacturing strategy is similar to the hierarchical view. At the highest stage, Stage 4, manufacturing is "externally supportive" of the company's objectives. Stage 4 firms actively involve manufacturing in marketing and engineering decisions, seek to anticipate new technological developments, and pursue long-range programs to acquire manufacturing capabilities before they are needed.

Many manufacturing firms remain at Stage 1 or Stage 2 unless they are forced by competitive pressures to move to Stage 3 or Stage 4. The transition from Stage 3 to Stage 4 is the most difficult. Hayes and Wheelwright state that Stage 4 companies are generally recognized by three characteristics: continual in-house process innovation, in-house development of manufacturing equipment, and close attention to both the structural and infra-structural aspects of manufacturing management [7]. These characteristics are frequently seen in the excellent manufacturing operations of German and Japanese firms.

The lessons of global competition can be concisely re-stated in terms of the Hayes-Wheelwright theory. At a minimum, the successful EC competitor must achieve a high degree of coordination across national boundaries and must systematically address production and distribution issues. This degree of strategic development corresponds to a Stage 3 strategic role for manufacturing. In industries where global cost leadership or differentiation becomes the prevailing strategy, even more will be required. Companies will be forced to develop in-house manufacturing engineering expertise and to measure themselves by global standards of cost and quality. This level of strategic development corresponds to Stage 4 in the Hayes-Wheelwright framework.

RECOMMENDATIONS FOR MANUFACTURING STRATEGY

From the preceding discussions, the major implications of the Single Market for manufacturing strategy can be identified. The major impacts from the single market include:

1. The EC will be an integrated market with 320 million consumers.

2. The harmonization of product standards will create numerous opportunities for product standardization and product re-design.

3. The elimination of tariffs and border controls will create opportunities to serve several EC countries from one source point and to identify new suppliers.

4. New patterns of distribution will emerge.

5. Global or EC-wide standards for measuring cost and quality performance will become dominant.

6. New organizational patterns must balance the benefits of central control and flexibility.

7. Competitive pressures will increase, and more Stage 3 and Stage 4 competitors will emerge.

Several authors have provided similar lists of typical decision categories related to manufacturing strategy [2], [5], [7]. One list, due to Hayes and Wheelwright, includes capacity, facilities, technology, suppliers and vertical integration, workforce, quality, production planning and materials control, and organization [7]. Table 1 summarizes the implications of the Single Market program for each of these decision categories. The recommendations in Table 1 are based on previous research in manufacturing strategy [2], [5], [7], [10], [11] and on the preceding discussion of global competition.

Table 1: Implications for Manufacturing Strategy

Manufacturing Strategy Area	1992 Impact(s)	Manufacturing Strategy Implications
Capacity	Large integrated market	Develop capacity strategy which addresses capacity requirements throughout the EC.
	Opportunity to serve several EC countries from one plant and dominance of global cost standards	Build capacity in larger increments or expand existing facilities to realize economies of scale.
	Product standardization	Convert capacity to new products.
Facilities	Opportunity to serve several EC countries from one plant.	Determine facility locations and missions on EC-wide basis. Consolidate manufacturing operations to realize economies of scale.
	New patterns of distribution.	Evaluate impact of new distribution patterns on facility location.
	Dominance of global cost standards.	Locate facilities to obtain cost advantages in labor, materials, or transportation. Evaluate the need to upgrade manufacturing methods, remove bottlenecks, and improve materials handling in existing facilities. Evaluate need for changes in facility focus.

Table 1: Implications for Manufacturing Strategy (Continued)

Manufacturing Strategy Area	1992 Impact(s)	Manufacturing Strategy Implications
Technology	Large integrated market and greater product standardization	Re-design products and processes. Standardize components where possible to simplify coordination and realize economies of scale.
	Dominance of global cost and quality standards.	Move down product-process matrix toward more efficient technology. Systematically evaluate new process technology and automation technologies such as CAD/CAM and FMS. Share technological advances among affiliates.
	Probable emergence of Stage 3 or Stage 4 competitors.	Address long-term manufacturing trends systematically. Evaluate need to upgrade Manufacturing Engineering capability.
Suppliers and Vertical Integration	Greater product standardization.	Re-evaluate make/buy decisions. Evaluate opportunities for vertical integration. Negotiate volume contracts.
	New distribution patterns.	Evaluate potential new suppliers and negotiate contracts as appropriate. Take advantage of opportunities to reduce inbound and outbound freight costs.
	Dominance of global cost and quality standards.	Evaluate benefits of sole sourcing vs. multiple sourcing. Initiate vendor certification or qualification. Consider need for Just-in-Time delivery.
Workforce	Large integrated market and opportunity to serve several EC countries from one plant.	Evaluate policies concerning wage payment, hiring, and employment stability.
	Dominance of global cost and quality standards.	Re-train workers as required by new manufacturing methods and technologies. Hire additional skilled workers as needed.

Table 1: Implications for Manufacturing Strategy (Continued)

Manufacturing Strategy Area	1992 Impact(s)	Manufacturing Strategy Implications
Quality	Dominance of global cost and quality standards.	Implement Total Quality Control. Certify vendors and consider eliminating incoming inspection.
	Probable emergence of Stage 3 or Stage 4 competitors.	Use quality as a competitive weapon. Stress worker participation, continual improvement, and technical competence in management.
Production Planning and Materials Control	Opportunity to serve several countries from one EC location and new patterns of distribution.	Coordinate aggregate production planning and distribution among facilities. Set new stock levels as required. Modify production planning and scheduling systems in accordance with new distribution patterns.
	Dominance of global cost and quality standards.	Develop "error-free" inventory control systems. Modify materials handling to achieve maximum efficiency. Consider Just-in-Time inventory control for repetitive manufacturing.
Organization	New organizational patterns.	Centralize control of strategy and insure adequate coordination among subsidiaries. Allow subsidiaries enough autonomy to insure effectiveness. Facilitate sharing of technology among plants and subsidiaries.

Some manufacturing strategy options will still be constrained by government policy. For instance, laws relating to employment stability and plant closings are much more stringent in parts of the EC than in the United States. Consequently, rationalizing capacity will be difficult for some firms. It is possible that political and economic pressures will impede major aspects of the Single Market program, such as the harmonization of product standards or further deregulation of transportation. In the final analysis, the extent to which the Single Market becomes a reality will depend on the commitment of EC leaders.

CONCLUSIONS

Which firms are most likely to benefit from the Single Market? The most competitive firms will be those which have a global or EC-wide corporate strategy and have attained Stage 4 competence in manufacturing strategy. These firms fully recognize the potential of manufacturing as a competitive weapon, have achieved superior integration of manufacturing strategies with marketing and engineering strategies, are accustomed to dealing with new technologies and practices in a proactive manner, and continually improve their products, processes, and infrastructure. These capabilities give Stage 4 firms the flexibility to adapt smoothly to new conditions, such as the Single Market.

In some industries, Stage 3 firms with an EC-wide strategy may also be competitive. These companies have a history of pursuing manufacturing strategy objectives that are in harmony with corporate and business unit objectives. Stage 3 firms also have the skills to manage changes in manufacturing technology and practice.

Many firms which have pursued protected market or national responsiveness strategies face a difficult competitive situation. These firms may now find themselves competing with manufacturers who meet global standards of excellence. The less efficient firms will need to adopt new corporate strategies and also to upgrade their manufacturing strategy capabilities from Stage 1 or Stage 2 to Stage 3 or Stage 4. Thus, the Single Market will encourage the improvement, consolidation, or merger of inefficient operations.

Companies which export from the United States to the EC will now face a uniform EC customs duty and product harmonization standards which could be tailored to the current production capabilities of European firms. A number of analysts have suggested that American manufacturers face a "Fortress Europe" where the uniform customs duty and the new product standards will be substantial barriers to American exporters. American companies should respond to this possibility in a proactive manner. Specifically, exporters should consider the following actions: evaluating their future competitive position in EC markets, revising corporate and manufacturing strategies accordingly, modifying their production systems as required, and participating in trade association efforts to ensure that the new product standards do not discriminate against American exporters. Joint ventures with EC firms or the location of manufacturing facilities within the EC may be desirable in some cases.

Previously, competitive pressures have forced many American firms to upgrade to manufacturing's strategic role to Stage 3 or Stage 4. However, some firms which export to the EC and many European companies have not previously faced such pressures. The competitive pressures of the Single Market will force these firms to meet global standards for cost and quality performance. These firms would be well-advised to assess their future competitive position carefully and consider upgrading manufacturing's strategic role to Stage 3 and perhaps eventually to Stage 4.

REFERENCES

1. Bartlett, C. A., "Building and Managing the Transnational: The New Organizational Challenge". In M. E. Porter (Ed.), Competition in Global Industries, (pp. 367-401). Boston, Massachusetts: Harvard Business School Press, 1986.

2. Buffa, E. S. Meeting the Competitive Challenge: Manufactur-turing as a Competitive Weapon. Homewood, Illinois: Dow Jones-Irwin, 1984.

3. Cvar, M. R. "Case Studies in Global Industries: Patterns of Success and Failure". In M. E. Porter (Ed.), Competition in Global Industries, (pp. 483-515). Boston, Massachusetts: Harvard Business School Press, 1986.

4. Flaherty, M. T., "Coordinating International Manufacturing and Technology". In M. E. Porter (Ed.), Competition in Global Industries. Boston, Massachusetts: Harvard Business School Press, 1986.

5. Gunn, T. G. Manufacturing for Competitive Advantage: Becoming a World Class Manufacturer. Cambridge, Massachusetts: Ballinger, 1987.

6. Hamel, G. and C. K. Prahalad, "Do You Really Have a Global Strategy?", Harvard Business Review, (July-August 1985), 139-148.

7. Hayes, R. H. and S. C. Wheelwright, Restoring our Competitive Edge: Competing through Manufacturing. New York: John Wiley and Sons, 1984.

8. Hout, T., M. E. Porter, and E. Rudden, "How Global Companies Win Out", Harvard Business Review, (September-October 1982), 98-108.

9. Porter, M. E., "Competition in Global Industries: A Conceptual Framework". In M. E. Porter (Ed.), Competition in Global Industries. Boston, Massachusetts: Harvard Business School Press, 1986.

10. Skinner, W., "Getting Physical: New Strategic Leverage from Operations", Journal of Business Strategy 3, (Spring 1983), pp. 74-79.

11. Starr, M. K., "Global Production and Operations Strategy", Columbia Journal of World Business 19, no. 4 (Winter 1984), 17-22.

12. Swamidass, P. M. and W. T. Newell, "Manufacturing Strategy, Environmental Uncertainty, and Performance: A Path Analytic Model", Management Science 33, no. 4 (April 1987), 509-524.

13. Tully, S., "Europe Gets Ready for 1992", Fortune, 117 (February 1, 1988), 81-84.

14. United Kingdom Department of Trade and Industry and Central Office of Information, "The Single Market: The Facts", London: HMSO, 1989.

4

Understanding Functional Decision Making at the Marketing/Manufacturing Interface

Victoria L. Crittenden, Boston College[1]

Abstract

Traditionally, each functional area performs specialized portions of the organization's tasks. Practitioners and academics tend to tackle such "vertical activities" more easily, since the activities require expertise in only one functional area. Horizontal activities, however, cut across functional areas and require much more coordination among functions. The research reported here examines horizontal activities by utilizing a simulation model, PROD/ACCT, to examine profit outcomes of various functionally-rooted decision rules. The results show the potential for such a decision support tool, as well as providing some interesting suggestions regarding constrained and excess capacity.

Introduction

While potential conflict exists between any of two or more business functions within an organization, Hayes and Wheelwright (1984 p. 199) contend that, "the marketing/manufacturing interface is the focal point of much more frequent and heated disagreement than occurs between other pairs of functions." This is hardly surprising given the often conflicting goals of the two functions (cf. Shapiro 1977; Wind 1981). Broadly, marketing's goal structure focuses upon demand stimulation while, conversely, manufacturing's goal structure centers around supply regulation.

Shapiro (1987b) contends that the source of the marketing/manufacturing conflict lies with account and product mix decisions. Account mix and product mix tend to pose as constraints on demand and supply decisions, respectively. From a marketing perspective, selecting which accounts to serve determines volume and variety of required supply. From a manufacturing perspective, however, determining the optimal mix of products to produce confines demand fulfillment to available product volume.

This research approaches the marketing/manufacturing interface using a product/account framework. The variables and relationships of interest do not lend themselves, easily, to analytic solutions. As such, we employ simulation techniques. The simulation provides a laboratory for experimentation and

---

[1]The author acknowledges fellowships from the American Association of University Women, the Stephen X. Doyle Doctoral Fellowship at the Harvard Business School, and the Boston College, CSOM, new faculty grant program as partial funding for this research. The author also acknowledges Professors Robert J. Dolan, Benson P. Shapiro, and Patrick J. Kaufmann for their comments and support in this research endeavor.

analysis of problems that cannot be solved by other means (Lee and Moore 1975).

The paper is organized as follows. The next section provides a brief overview of four mechanisms for facilitating interfunctional coordination. Then, we describe the simulation model. The third section presents the results of evaluating the utility of a decision support/simulation model and analyzing various functional-level decision rules. The paper concludes with a brief discussion of positive and negative aspects of using such a model.

Facilitating Interfunctional Coordination

Exhortations and predictive frameworks suggest four major "facilitating mechanisms" for improved interfunctional coordination: (i) organizational design, (ii) communication, (iii) evaluation systems, and (iv) models.

Many academicians and firms support the need for an *organizational design* conducive to nurturing integrative activities and possibly the demise of older, functionally specialized structures (cf. Goldhar and Jelinek 1983). Design suggestions vary and include mixed structures (Boston Consulting Group 1985), decentralization of authority (Ruekert, Walker, and Roering 1985), teams and transplanting people into various functional roles for a limited time (Carroad and Carroad 1982), and matrix organizations (McDonough 1984). Firms experimenting with design variations include: Deere & Co.'s Horicon Works, Honeywell Inc.'s Building Controls Division in Golden Valley, Minnesota, and Ford Motor Company with its Taurus-Sable project.

Breakdowns in *communication* are an oft-cited reason for a lack of integration among functional areas (cf. Gupta, Raj, and Wilemon 1985; Millman 1982; New and Schlacter 1979), and several writers espouse the need for improved communications as a requisite for interfunctional effectiveness (e.g., Clare and Sanford 1984; Kiel 1984; O'Keefe and Chakrabarti 1981). 3M, Wright Line, and National Starch & Chemical are three firms that report using improved communication efforts to integrate functions.

A conceptual model (Carroad and Carroad 1982) and results from measuring concerns about R&D and marketing interactions (Souder 1981; Young 1979) suggest that *reward systems* should reflect concerns of both parties in an internal exchange relationship. Unfortunately, it is rare to find reports of how reward measures are implemented in practice.

Academically, there is little research *modeling* the interface between functional groups. Jackson and Shapiro (1979) outline the basic approach product managers can take to put together an industrial product line computer model. While other academics suggest using models as interface tools, the results of their attempts are models too complex for implementation by most managers (e.g., Anderson and Taylor 1977) or interesting from a theoretical view yet provide no means for usage (e.g., Showalter and White 1986). However, interfunctional models are becoming more and more common in the business community (e.g., AT&T Bell Laboratories, American Airlines, The Christian Brothers).

This research focuses upon the use of models to improve interactions between marketing and manufacturing. We suggest a joint marketing/manufacturing model (PROD/ACCT) which incorporates two major issues at the interfunctional interface: (i) both marketing and manufacturing concerns (i.e., sales volume, price, cost, and capacity) are components of the model and (ii) demand and supply regulation comes about through the evaluation of functionally-oriented decision rules concerning what to produce, for whom, and when to produce the product.

PROD/ACCT Framework and Simulation Model

A truly integrative model of marketing and manufacturing allows for both supply and demand concerns. Such a micro model considers market demand by segmenting to the individual customer level and manufacturing supply decisions by evaluating individual products relative to customers the firm selects to serve:

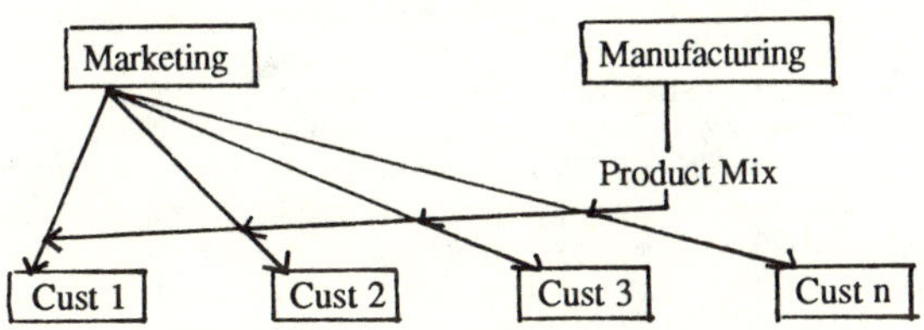

Shapiro (1987a) captures, conceptually, the two major dimensions of supply and demand in a product/account matrix (Figure 1). This current research translates Shapiro's conceptual model into a decision model (PROD/ACCT) which allows us to evaluate rigorously the impact of marketing- and manufacturing-related decisions on firm profitability. [We define profitability as product revenue less marketing- and manufacturing-related costs. We do not include overhead in this measure.]

Model Components

Individual cells in the PROD/ACCT simulation model include various marketing and manufacturing concerns. Capacity and costs encompass manufacturing supply concerns within each product cell. These include: (i) available production capacity, (ii) variable cost of production, (iii) setup cost of production (discretionary fixed cost), and (iv) holding cost (indirect variable cost). Sales volume and price comprise marketing demand concerns. These include: (i) size of individual orders, (ii) expected sales over planning period, (iii) customization requirements, and (iv) product price per account. An additional marketing demand variable arises in the rows of the matrix -- that of cost to serve each account. The production capacity component presents the major constraint in the model. That is, manufacturing cannot produce more, in the short-term, than capacity allows.

Functional-Level Decision Rules

Over time, management practitioners and academics have accumulated a long list of rules for making product and account mix decision. Essentially, these rules specify how to allocate capacity. Capacity allocation decisions result in differing outcomes in the cells of the product/account matrix.

Reviewing decision rules in the literature (e.g., Chase and Aquilano 1989) and discussing procedures with marketing and manufacturing managers lead to consideration of 18 possible priority/decision rules for making account and product mix decisions. The initial eight are relevant for constrained production capacity situations. The other ten for excess capacity situations.

The 18 rules are:

<u>Constrained Production Capacity</u>

1. Fill orders relative to desired amount
2. Fill orders as proportion of total expected sales
3. Fill best prices first
4. Fill orders by account priority
5. Fill orders based on account profitability
6. Fill most profitable account/product cell first
7. Fill certain minimum amount of each account
8. Produce lowest cost products

<u>Excess Production Capacity</u>

9. Produce what manufacturing "likes to make"
10. Produce product most consistent with where excess capacity occurs
11. Produce product where least amount of excess capacity occurs
12. Produce product with least customization requests
13. Produce product that can sell easily
14. Produce most profitable product
15. Produce product with most stable demand
16. Produce product that may need next quarter
17. Produce product that is most often out-of-stock
18. Leave capacity idle

Model Manipulation

The software we use to operationalize PROD/ACCT is Interactive Financial Planning System (IFPS) by EXECUCOM System Corporation. IFPS incorporates a "what-if" function that allows profit-impact evaluation of the various priority/decision rules.

PROD/ACCT consists of three major modules: (i) demand, (ii) marketing, and (iii) manufacturing. Figure 2 portrays the interactions of the three modules. Priority/decision rules tell the model what to do when capacity is less than (i.e., evaluate using the constrained rules) or greater than (i.e., evaluate using excess rules) demand requirements. Thus, the key item in the model is the difference between production hours available to make each product (i.e., model constraint) and production hours needed to meet demand for each product. Negative values indicate constrained capacity and positive values indicate excess capacity.

For each constrained or excess priority/decision rule, the model provides company profit implications by subtracting both marketing and manufacturing costs from sales revenue. Utilizing the product/account framework, the model also provides profitability information for each account, product, and product by account.

PROD/ACCT Simulation Results

Using firm data, we fulfill two objectives of this research: (i) evaluate the utility of a decision support model for improving interfunctional decision-making and (ii) take the simulation beyond particulars of the data and algorithm to determine if there are rule outcome consistencies across various account mix, demand, and margin scenarios.

Model Utility

Data from two companies are used in this initial investigation. One set of data comes from a firm which produces six major paper/plastic cups and dinnerware products. The company has around 500 customers in seven geographic segments. The company experiences constrained production capacity during the first two quarters in its paper conversion process and during all quarters in its thermoforming cups and lids and thermoforming dinnerware processes. The firm is faced with three major decisions: (i) customizing products for one of its major accounts, (ii) accepting the order of a potentially large account (an account which the company has been pursuing heavily), and (iii) decreasing price during the off season.

Using PROD/ACCT to evaluate these three options, we find these situation-specific outcomes: (i) the customization of products for a major account yields long-term profits, given the possibility that the firm might lose the account otherwise, (ii) sacrificing accounts to obtain one large account has a positive impact upon profit, and (iii) maintaining idle capacity yields greater profits than decreasing prices.

The second set of data is from a firm experiencing long-term excess capacity. The company engages in the manufacture and sale of electronic analog circuit modules, isolation amplifiers, and power converters. The company categorizes its customers according to first, second, and third tier accounts. The firm is curious about the outcomes of two possible alternatives: (i) increasing rep commissions on sales of the best selling product and (ii) increasing rep commissions, as well as, decreasing price on the best selling product. An implicit assumption in these alternatives is that increased selling effort (a nonprice component of the marketing mix) shifts the product's demand curve. A second assumption, when decreasing price, is that product demand is elastic.

Both alternative scenarios result in higher profits than possible under the firm's current situation. The major drawback to both alternatives is that the firm bases sales estimates on production capacity rather than realistic demand expectations.

The results of analyzing company alternatives, using PROD/ACCT, supports the notion that a simulation model enhances product and customer mix decisions. In practice, PROD/ACCT allows a firm to quickly and easily see potential outcomes of various decisions. Additionally, this initial investigation validates the utility of the model in terms of fit with marketing and manufacturing behavior (e.g., representing a real marketing/manufacturing dilemma) and fit with organization behavior (e.g., decision rules compatible with the way functional managers make decisions).

Priority/Decision Rule Outcome Consistency

To achieve the second objective of this research, we present two groups of account mix scenarios: (i) many accounts with small orders (Level 1), and (ii) few accounts with larger orders (Level 2). Within each of these groups, we evaluate two variables: (i) three demand levels (Demand 1, 2, and 3), and (ii) three margin levels (Margin 1, 2, and 3). Capacity is fixed in this set of simulation runs. This 2 x 3 x 3 design results in 18 different scenarios in which we evaluate decision making using the priority/decision rules.

1. Level 1/Demand 1/Margin 1
2. Level 1/Demand 1/Margin 2
3. Level 1/Demand 1/Margin 3

4. Level 1/Demand 2/Margin 1
5. Level 1/Demand 2/Margin 2
6. Level 1/Demand 2/Margin 3

7. Level 1/Demand 3/Margin 1
8. Level 1/Demand 3/Margin 2
9. Level 1/Demand 3/Margin 3

10. Level 2/Demand 1/Margin 1
11. Level 2/Demand 1/Margin 2
12. Level 2/Demand 1/Margin 3

13. Level 2/Demand 2/Margin 1
14. Level 2/Demand 2/Margin 2
15. Level 2/Demand 2/Margin 3

16. Level 2/Demand 3/Margin 1
17. Level 2/Demand 3/Margin 2
18. Level 2/Demand 3/Margin 3

We utilize data from the firm experiencing excess capacity in the initial investigation as the base case environment for the simulated results (Level2/Demand1/Margin1). Using this particular set of firm data flows us to monitor the move from excess to constrained capacity (i.e., as we double/triple demand), as well as compare priority/decision rule consistency.

Overall, we find strong correspondence among rule rankings in the simulation scenarios. Any anomalies are explained away by firm-data insight. The results of the 18 simulation scenarios suggest the following:

1. Thresholds emerging in the data suggest that extent of demand, rather than account mix or margin, should be the critical component of the functional decision-making process. Only when demand is tripled do we see focusing upon account mix (i.e., few accounts) as important. Additionally, the cost of using different rules increases rather dramatically as demand increases.

Regarding excess capacity:

2. Evaluating various generally-accepted procedures for allocating capacity, the greater dispersion of profits for constrained production capacity supports what some managers believe, but have been unable to show, about capacity allocation. That is, the allocation of constrained capacity is more critical than the allocation of excess production capacity. Experiencing excess capacity avoids problems of product mix and account selection.

3. Such a conclusion implies, also, that the allocation of constrained capacity is a functional strategy decision, but that the management of excess capacity is a corporate strategy decision. That is, under conditions of constrained capacity, functional managers can effect firm profitability through decisions concerning allocation of constrained capacity. However, the lack of dispersion among priority/decision rules for excess capacity suggests little impact at the functional level and the need for corporate level involvement in order to decide on such issues as firm objectives (e.g., profit maximization vs. steady employment) and diversification.

Regarding constrained capacity:

4. Producing the highest price product dominated in all scenarios. This short-term rule dominated all other short-term rules by at least 20 percent. Interestingly, this rule ignores account relationships and focuses vertically within the product/account matrix by selling to both accounts and base customers. Caution must be used in extrapolating this finding, however. This result may be confounded by the data used in the analysis. That is, the highest priced product was also a very high margin product.

5. A company has to evaluate the cost of account relationships. For example, a 20 percent cost at Demand 2 may be a low price to pay to have the security of account relationships. However, the 30 to 40 percent cost to serve at Demand 3 may be too high a price to pay for such security.

6. Subjective account priority rules do very poorly. Rule 4 is a subjective rule in that managers rank accounts on criteria that may not be substantiated quantitatively (e.g., closer geographically and thus easier to contact, good working relationship with purchasing agents). Profit outcome of Rule 3 is greater than Rule 4 by anywhere from 31 to 70 percent. This suggests the need for strong justification for account rankings, rather than just intuition or "gut feel."

7. Short-term rules fare better than long-term rules. However, this may be an artifact of the model. The current analysis does not extend far enough into the future to do justice to long-term oriented rules since these various simulation runs do not include changes if the firm loses specific accounts. Such a long-term orientation, using a simulation model, suggests a need for a probabilistic method of forecasting account demand in order to

simulate best and worst cases in time t+n. In this way, we would be better able to compare rules over the long-term. However, we can safely say that it will be difficult to justify a long-term orientation when we see anywhere from 20 to 46 percent difference in short-term profitability potential.

Summary

While firm data and systematic variation of the data support strongly conclusions regarding excess and constrained capacity decision making, this research is not without limitations, however. Foremost, the PROD/ACCT simulation model provides only comparative results. We are unable to suggest an optimal capacity allocation procedure, but show conditions under which we expect certain outcomes to occur. However, the model with its 18 trial and error simulations (i.e., 18 priority/decision rules) is, in all likelihood, capturing the optimal allocation procedure. Additionally, the simulation format appears to be a usable managerial tool.

From a broader perspective, there is concern as to whether the model represents, honestly, interfunctional relations between marketing and manufacturing. That is, can we quantify such subjective interactions? We recognize that the simulation model is an abstraction of reality. However, incurring the costs of quantifying subjective interactions allows the incorporation of situational aspects of marketing and manufacturing interactions which we were unable to do prior to the model's development. Using the model as a "means" of understanding the possible consequences of individual and joint marketing and manufacturing decision-making shows rather strongly that decision criteria functionally-anchored likely give poorer results than cross-functional decision criteria. Additionally, the development, operationalization, and discussion of the model should lead to positive interfunctional discussions rather than negative confrontation.

FIGURE 1

Product/Account Profitability Framework

MANUFACTURING SUPPLY DECISIONS

		Product 1	Product 2	Product j	Account Profit
	Account A	$\Pi A1$	$\Pi A2$	ΠAj	ΠA
MARKETING DEMAND DECISIONS	Account B	$\Pi B1$	$\Pi B2$	ΠBj	ΠB
	Account i	$\Pi i1$	$\Pi i2$	Πij	Πi
	Base Sales	$\Pi Base1$	$\Pi Base2$	$\Pi Basej$	$\Pi Base$
	Product Profit	$\Pi 1$	$\Pi 2$	Πj	Firm Profit

FIGURE 2

Flow Diagram of PROD/ACCT Simulation Model

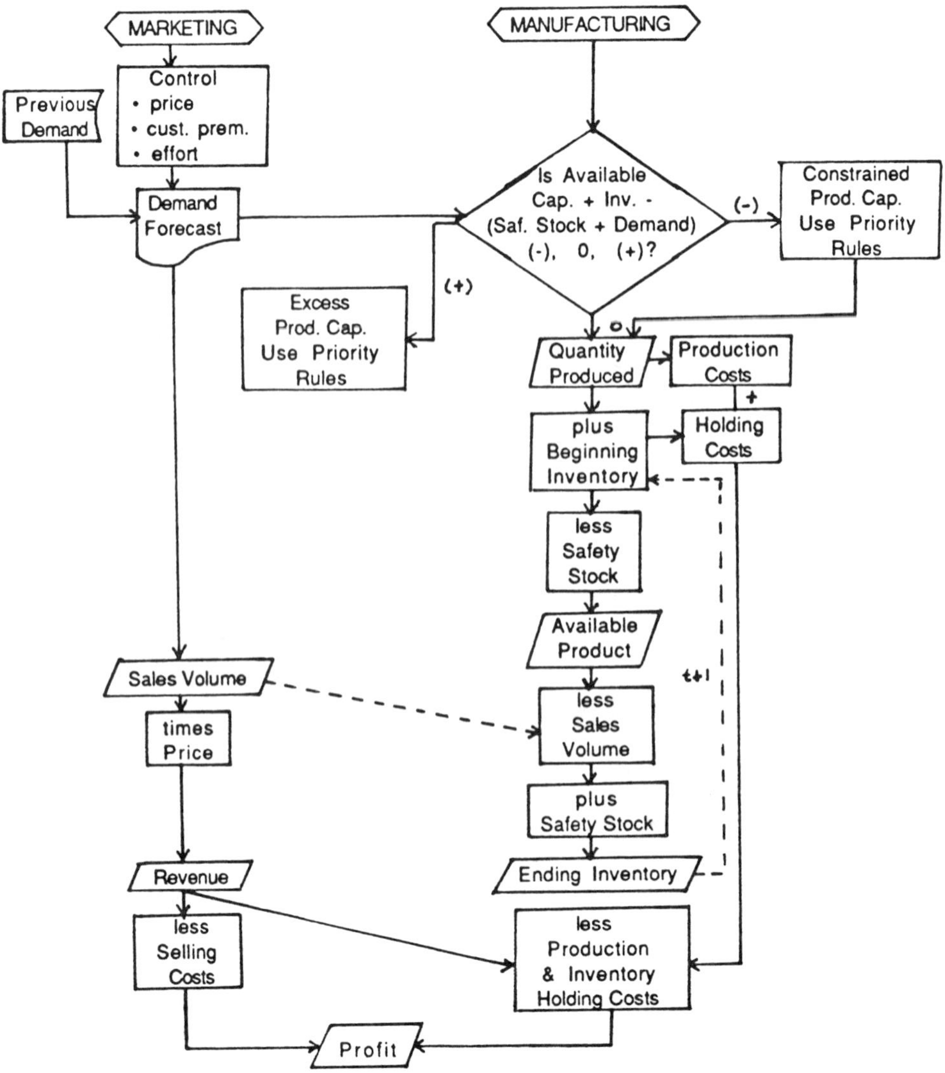

REFERENCES

Anderson, Paul F. and Bernard Taylor (1977). "Marketing/Quality Control Interface: The GERT Approach." Industrial Marketing Management, (6):420-428.

Boston Consulting Group (1985). Organizing for Electronics. Boston, MA: BCG.

Carroad, Paul A. and Connie A. Carroad (1982). "Strategic Interfacing of R&D and Marketing." Research Management, (January):28-33.

Chase, Richard B. and Nicholas J. Aquilano (1989). Production and Operations Management: A Life Cycle Approach, 5th edition. Homewood, IL: Richard D. Irwin, Inc.

Clare, Donald A. and Donald G. Sanford (1984). "Cooperation and Conflict BetweenIndustrial Sales and Production." Industrial Marketing Management, (13): 163-169.

Goldhar, Joel D. and Mariann Jelinek (1983). "Plan for Economies of Scope." Harvard Business Review, (November-December):141-148.

Gupta, Ashok K., S. P. Raj, and David Wilemon (1986). "A Model for StudyingR&D-Marketing Interface in the Product Innovation Process." Journal of Marketing, (April):7-17.

Hayes, Robert H. and Steven C. Wheelwright (1984). Restoring Our Competitive Edge: Competing Through Manufacturing. New York, NY: John Wiley & Sons.

Jackson, Barbara B. and Benson P. Shapiro (1979). "New Way to Make Product Line Decisions." Harvard Business Review, (May-June):139-149.

Kiel, Geoffrey (1984). "Technology and Marketing: The Magic Mix?" Business Horizons, (May-June):7-14.

Lee, Sang M. and Laurence J. Moore (1975). Introduction to Decision Sciences. New York: Petrocelli/Charter.

McDonough, Edward F. III (1984). "Needed: An Expanded HRM Role to Bridge the Gap Between R&D and Manufacturing." Personnel, (May-June):47-53.

Millman, A. F. (1982). "Understanding Barriers to Product Innovation at the R&D: Marketing Interface." Management Decision, 20, p. 10-22.

New, D. E. and J. L. Schlacter (1979). "Abandon Bad R&D Projects with Earlier Marketing Appraisals." Industrial Marketing Management, (November): 274-280.

O'Keefe, Robert D. and Alok K. Chakrabarti (1981). "Coordination and Communication in Industrial Innovation: The R&D/Marketing Interface Problem." Baylor Business Studies, 12, pp. 35-43.

Ruekert, Robert W., Orville C. Walker, Jr., and Kenneth J. Roering (1985). "The Organization of Marketing Activities: A Contingency Theory of Structure and Performance." Journal of Marketing, (Winter):13-25.

Shapiro, Benson P. (1977). "Can Marketing and Manufacturing Coexist?" Harvard Business Review, (September-October):104-114.

Shapiro, Benson P. (1987a). "The Magic Matrix: Product and Accounts." Textual Note, Harvard Business School.

Shapiro, Benson P. (1987b). "The New Intimacy." Textual Note, #0-587-121, Harvard Business School.

Showalter, Michael J. and J. Dennis White (1986). "An Integrated Model for Demand-Output Management in Service Organizations: Implications for Future Research." Working Paper, Florida State University.

Souder, William E. (1981). "Disharmony Between R&D and Marketing." Industrial Marketing Management, (February):67-73.

Wind, Yoram (1981). "Marketing and the Other Business Functions." Research in Marketing, Vol. 5, pp. 237-264.

Young, H. Clifton (1979). "Effective Management of Research-Market Teams." Research Management, (March):7-12.

5

THE NATURE AND DYNAMICS OF TECHNOLOGICAL CHANGES IN AN AUTOMOBILE TIRE COMPANY - A CASE STUDY

C. Carl Pegels, Ph.D.
James R. Meindl, Ph.D.
John M. Thomas, Ph.D.
School of Management
SUNY/Buffalo
Buffalo, NY 14260

ABSTRACT

In this paper we describe how technological changes in an American automobile tire manufacturing firm helped to improve productivity and quality to a considerable extent over only an 18 month period of time. Technological changes considered not only include higher level technology machinery but also process and numerous managerial changes plus extensive retraining of the work force.

INTRODUCTION

This is a study of technological change which occurred at an American tire manufacturer. The American tire manufacturer, Dunlop Tire Corporation, had delayed technological changes in its manufacturing plants because of a shortage of capital and by concentrating on the replacement tire market. The replacement tire market places much lower demand on the technological level of a tire manufacturer than does the original equipment market. The original equipment market for tires places a discipline on tire manufactures and it is generally believed by tire industry executives that to be successful in the tire industry a tire manufacturer must be able to compete in the original equipment market for tires.

With the globalization of the tire market, competition in the tire industry has increased significantly. As a result it has become increasingly difficult to compete in the tire market, especially for the smaller, low technology tire manufacturers with their lower levels of productivity.

Dunlop Tire Corporation, the American manufacturer of Dunlop tires and a couple of secondary brand tires, had become a marginal, low technology and largely replacement tire producer in the mid 1980s. To avoid considerable shrinkage in their operations and possible bankruptcy, it agreed to be acquired by Sumitomo Rubber Industries (SRI) in 1986. SRI decided to retain the previous American management team and to introduce technological change in the American firm through the use of Japanese technical advisors and through a massive infusion of capital. The capital was needed to improve the technological level of Dunlop so that it could profitably compete not only in the replacement tire market but also in the much more competitive original equipment tire market.

This paper presents a case study of how SRI was able to restructure Dunlop Tire Corporation from a low technology tire manufacturer to a high technology tire manufacturer.

Sumitomo Rubber Industries had a considerable original equipment tire market in Japan. As Japanese automobile manufacturers began shifting some of their assembly operations to the United States, SRI wanted to maintain its original equipment tire market share in the United States market. Through the acquisition of Dunlop Tire Corporation, SRI was able to achieve its goal of maintaining their original equipment tire market share.

Below we shall discuss in detail the two types of technological changes which were introduced by SRI in the Dunlop organization. The two types of technological changes are manufacturing process technological change and product technological change. Both types of technological change were needed by Dunlop to become competitive in the original equipment tire industry and to remain competitive in the replacement tire market in the United States.

TWO TYPES OF TECHNOLOGICAL CHANGE

In the tire manufacturing industry, as well as in most manufacturing industries, we can find two major areas of technological change. They are technological change in the manufacturing process and technological change in the manufactured product. The two technological changes are frequently related as for example in the case where a technological change in the manufacturing process is necessary in order to achieve a technological change in the product. In the Dunlop case this situation arose when Sumitomo Rubber Industries discovered that to achieve the technological, performance and quality levels of tires for the original equipment market would require numerous technological changes in both tire building machinery, in raw material mixing and in the tire curing process.

TECHNOLOGICAL CHANGE IN THE MANUFACTURING PROCESS

The technological change in the manufacturing process can be further subdivided into:

1. up-to-date high technology equipment
2. up-to-date high technology machinery
3. restructured lower and middle management practices style
4. improved worker communication and participation
5. retaining of workers, of lower management and of middle management

To respond to the high quality and unique specification requirements of original equipment manufacturers, the tire manufacturers need the most up-to-date equipment in terms of electronic technology, chemical testing and other demands. Since original equipment customers, in this case the automobile, motorcycle and truck manufacturers are in essence able to dictate their specifications and quality levels to the tire manufacturers the tire manufacturers must respond to the original equipment customers by having installed the most up-to-date equipment in their tire plants. Level of equipment technology is therefore determined by competitive necessity. If a tire manufacturer wants to compete in the original equipment tire market he must be willing and able to have the requisite high technology equipment available in his plant.

Up-to-date high technology machinery to manufacture tires is similarly dictated by original equipment customers. Again it is impossible for a tire manufacturer to be in the original equipment market unless he is able to meet the original equipment customer's specification and quality requirements. These specifications and requirements can only be met with the latest technology machinery. There is another reason why the latest technology machinery is critical to the success of a tire manufacturer - that reason is cost competitiveness. The profit margins on original equipment tires are very thin. Hence, original equipment tires can only contribute to the tire manufacturer's profits through the utilization of highly efficient manufacturing processes.

Restructured lower and middle management practices and the resultant management style may not be recognized by everyone as a technological change. It is true that the management restructuring process does not require the acquisition of new physical technology. However, the improvements in effectiveness and efficiency that have been wrought by a restructured management approach have been significant and often critical in the survival of a firm. The restructuring process centers on improved two-way communication with workers, and on utilizing worker skills and knowledge about their jobs through quality circle and other worker utilization activities.

Above we described the lower and middle management need to recognize and utilize worker communication and participation in the management process. To achieve restructuring of lower and middle management also requires retraining of the worker to view himself as a more comprehensive participant in the manufacturing process than in the past. Worker communication and participation can only be achieved with the full cooperation of the workers. Workers must view themselves as important contributors and be active as contributors to the overall manufacturing process. In the past the worker usually was concerned with his own narrow environment and received little feedback about what occurred outside of his own area. To achieve worker participation and to improve worker communication is not easy to achieve and requires the total cooperation of both workers and lower and middle level management as well as the support of top management.

Retraining of workers, lower management and middle management is necessary to achieve the integration of the four technological changes described above. New technology in the form of up-to-date machinery and equipment can only be acquired at considerable cost. The same is true for technological changes achieved through restructured lower and middle management and improved worker communication and participation. To achieve the maximum benefit from all new technology introduced requires that the users of the new technology are properly trained to gain maximum benefit from the new technology. The term retraining is somewhat of a misnomer. Successful utilization of any new technology requires a continuing process of training and refresher training. Hence, training can be viewed as a technological change by itself.

Above we have described the technological change process in the tire manufacturing process at Dunlop following the acquisition of Dunlop by Sumitomo Rubber Industries. The conditions, the technology acquisition, the management restructuring and the worker and management retraining described above took place at Dunlop. The changes that took place as well as the resultant performance was costly to achieve but paid off quickly in improved quality and productivity in order to compete in the highly competitive original equipment international tire market.

Prior to the takeover of Dunlop Tire Corporation by Sumitomo Rubber Industries, Dunlop was largely a replacement tire manufacturer. The replacement tire market is quite competitive on price basis, but not very competitive on a quality and specification-oriented basis. The original equipment tire market on the other hand was very competitive on price basis, on quality basis and on specification basis. A replacement tire manufacturer is therefore able to remain in business as long as he can compete on a price basis and is able to meet the much lower quality and specification standards of that market. However, competition in the original equipment tire market requires that a tire manufacturer is able to compete on the three dimensions of price, quality and specifications.

In order to be competitive in both the replacement and the original equipment tire markets dictated to Dunlop that they introduce technological change to improve their competitive position in terms of price, quality and specifications.

Dunlop Tire Corporation attacked the low technological level of their product, tires, by manufacturing process improvement as described above and by considerably increasing their product research, product development and product testing manpower and facilities. Although, the short-term benefits to the increased resources devoted to research, development and testing were non-existent from a profit-ability point of view, the longer term benefits for Dunlop were considered to be substantial. They allowed Dunlop to be a significant player in the original equipment tire market. Long term plans called for Dunlop to devote about 40 percent of their tire output to the original equipment market and 60 percent to the replacement market.

As was stated before the original equipment tire market is less profitable yet requires a much higher level of quality and ability to respond to tire specification changes. It is the firm belief of Sumitomo Rubber Industries that the only way a tire manufacturer can compete in the world tire market, and the American tire market, in the long run, is by having a significant part of its output sold to the original equipment tire market.

Original equipment tire customers of tire manufacturers usually have unique tire specifications for each of their vehicle models. For instance, a sporty automobile requires a different ride, and probably is of a different weight, than a van, a station wagon, a luxury automobile, etc. As a result automobile manufacturers will have quite exact specifications for a given automobile model. A tire manufacturer must be able to respond to these specifications quickly and exactly. To be able to do this requires a high manufacturing process technology level but also a high product technology level. The high product technology level can only be achieved by having a well-staffed research and development group with state-of-the-art testing facilities including test track and other high technology testing equipment and testing expertise.

BEHAVIORAL IMPACT OF TECHNOLOGICAL CHANGE AT DUNLOP

Not surprisingly the acquisition of Dunlop by Sumitomo Rubber Industries impacted on many Dunlop Tire Corporations entities including factory workers, all management levels of manufacturing management, management of research, development and testing, marketing management and on top management.

Factory workers are still the most pleased with the SRI takeover, especially those at one of the two Dunlop plants slated for closedown prior to the SRI takeover because of low technology and poor competitive position. The SRI takeover essentially saved their jobs and the technological changes described above have substantially contributed to the favorable response by the factory workers to the SRI takeover. At the other Dunlop plant, a somewhat more competitive plant, though still low in technology, the idea of the SRI takeover was greeted with apprehension. However, SRI was able to convince the workers that their long-term interests would benefit from the takeover and workers at this plant now also speak favorably about their plant's new owners.

The reason for the favorable response of workers can be found in several areas. First, the massive capital investment for new technology equipment and machinery is a positive sign for continuation of the business.

Second, the improvement in communication and worker participation adds a new and welcome addition to the job. Third, the insistence of clean and pleasant surroundings of the work place enjoy massive support from the workers.

All levels of manufacturing management strongly support the changes that have occurred in the manufacturing area, the area most affected by the SRI takeover. The favorable reaction of the workers to the SRI takeover is also reflected by manufacturing management responses. At the senior level there was strong support for the SRI takeover because the massive inflow of new technology equipment and machinery made their job much easier. In addition they were getting extensive support from technical advisors who were sent to Dunlop from Japan. Furthermore, during the transition phase from low technology to high technology the pressure for high productivity abated somewhat thus taking some of the pressure off the senior manufacturing managers. Lower and middle managers also generally praised the SRI takeover. Again, supervising a work force which is happier and more cooperative following the SRI takeover is easier than supervising unhappy workers.

Management responsible for product research, product development and product testing are also quite content with the SRI takeover of Dunlop because their operations have been extensively expanded.

Marketing management is one group of managers who are negatively affected by the SRI takeover. Because of the concentration on original equipment tires by SRI there are fewer replacement tires available to satisfy the replacement market. Hence, marketing management is forced to concentrate on keeping tire dealers and tire wholesalers happy with lower tire allocations than they would like. Although, this tire shortage condition is of a temporary nature, a few years, it is difficult for marketing management to accept that they are being allocated fewer tires for the more profitable replacement tire market than the original equipment tire customers receive.

The marketing management group has very little involvement with original equipment customers. Dunlop's original equipment customers are largely Japanese automobile companies with automobile assembly plants in the United States. Sales to the original equipment market in Dunlop's case is largely handled by the Japanese connections in Japan. Hence, Dunlop's marketing management is essentially a marketing group in charge of replacement tires. As a result the marketing group understandably feels that their short-term interests are negatively affected.

At the top management level there is a mixture of feelings about the SRI takeover. They are aware that, although they are partially independent in managing the company, they are also serving at the pleasure of SRI and can be replaced at any moment. Top management concern centers largely on the lack of control it can exercise over short term performance and short term profitability. The extensive technological changes that have occurred or are under way are intended to improve long term profitability. But American management traditionally has a short-term orientation and in this case has real difficulty in dealing with relatively poor performance in the short run.

PERFORMANCE IMPACTS OF TECHNOLOGICAL CHANGE AT DUNLOP

There are several dimensions by which performance can be measured prior to, during and after the takeover of Dunlop by SRI. The first dimension is total tire production. This measure is largely one of level of activity, although product mix could clearly affect it also. The second dimension is a more accurate productivity measure. It measures minutes of production time per tire. It is of course also affected by product mix with large tires taking more time than smaller tires. Another productivity measure is pounds of tire per manhour. It has the advantage of considering product mix and is therefore not biased towards small tire production as the other productivity measure is. However, the minutes per tire measure is an international measure and allows international comparison. Neither one of the two productivity measure take cost of manpower into account thus not providing good inter-company or inter-country comparisons.

Other performance measures are financial performance for specific periods during the transition phase. Quality levels achieved in the form of returns, rejects or complaints also provide a comparison of performance. Still another productivity performance measure is scrap generated.

We were not able to obtain information on all performance measures described above. However, Table 1 presents a comparison of tire production at the two Dunlop plants in the USA at different periods spanning the transition period during which the various technological changes occurred. As can be observed very little change in output occurred during the 18-months period covered except for some quarter to quarter variations. Table 2 presents information on the two productivity performance measures. As one can observe substantial productivity improvements occurred during the 18-months period, the period during which most technological changes occurred. Especially at plant B substantial improvements occurred in the minutes per tire (M/T) measure. At plant H, on the other hand little change occurred in minutes per tire. The measure pounds per manhour (#/M) improved substantially at both plant B and plant H.

Table 3 presents the financial performance of DTC during the transition phase. We had not expected to see improvements in financial performance because of the dislocations due to the various technological changes which occurred during this time period. A clear improvement occurred in the last quarter. But the poor performance in the other quarters must be attributed to the dislocations in operations which occurred because of the numerous technological changes being implemented.

Table 1 Tire Output at Dunlop Plants B and H (delivered to warehouse)

Quarterly Period Ending	Dunlop Plant B	Dunlop Plant H
9-30-87	100%	100%
12-31-87	117%	101%
3-31-88	124%	115%
6-30-88	116%	112%
9-30-88	99%	103%
12-31-88	112%	99%

Note: Outputs at each plant are given as a percent of quarterly period ending 9-30-87.

Table 2 Productivity Output Measuresat Dunlop Plants B and H
(in percent of period ending 9-30-87)

Quarterly Period Ending	Dunlop Plant B		Dunlop Plant H	
	M/T	#/M	M/T	#/M
9-30-87	100	100	100	100
12-31-87	97	98	104	101
3-31-88	92	103	105	106
6-30-88	83	106	103	108
9-30-88	79	102	99	109
12-31-88	75	112	99	109

Notes: M/T = Minutes per tire - a decrease indicates improvement
#/M = Pounds per manhour - an increase indicates improvement

Table 3 Financial Performance - Dunlop Tire Corporation

Quarterly Period Ending	Quarterly Revenues	Quarterly Profits
9-30-87	100	(0.98)
12-31-87	92	0.02
3-31-88	90	(4.41)
6-30-88	106	(3.74)
9-30-88	120	(2.27)
12-31-88	107	1.42

Notes:

1. Quarterly revenues are shown as a percent of period ending 9-30-87.
2. Quarterly profits are shown as a percent of revenues.

6

MANUFACTURING STRATEGY IN A VERY LOW VOLUME PRODUCTION SETTING: The Reemergence of the U. S. Space Transportation Industry

Christopher E. Getner

General Dynamics Space Systems Division
P.O. Box 85990, MZ 32-7151
San Diego, CA 92138-5990
(619) 573-8228

Michael D. Packer

General Dynamics Space Systems Division
P.O. Box 85990, MZ 92-8260
San Diego, CA 92138-5990
(619) 547-7930

ABSTRACT

This article explores the reemerging field of America's space industry. In particular, the unusual business environment is discussed as it relates to the specific areas of manufacturing strategy. Predominantly, the article explores the case of General Dynamics Space Systems Division and its initial efforts in the commercial launch service market -- a subject that the authors are very familiar with.

BACKGROUND

The current Expendable Launch Vehicle (ELV) industry in the United States is one which has undergone a significant upheaval within the last three years. The initiation of Shuttle operations in the early 1980's forced NASA to legislate out of existence the then existing launch fleet, in order to ensure the success and support of the Shuttle system. The reliability and cost effectiveness of this new system was intended to replace the "old" technology ELVs and NASA needed to ensure that demand was high enough for the Shuttle to support the anticipated flight rates. As a result, when the U. S. returned to an "assured access to space" strategy following the loss of the Challenger, the industry and infrastructure capability of the older ELVs had suffered from the better part of a decade of decay. Worse than mere inactivity, the industry had gone through seven years of production close out mentality, where very infrequent and specialized procurements of hardware were the standard.

Amplifying the challenge within the ELV industry is the "exotic", highly technical nature of the business. With the rise of the first high technology industries in the 1960's it was observed that " the technical aspects of the company's operations can no longer be delegated to the engineers" (Miller and Rogers, 1964, p. 2). What has developed in the launch industry is the fact that technical challenges not only impact internal functional relationships, but also have become a strong force in external marketing decisions. The application of advanced microelectronics for ELV avionics offers the customer higher accuracy of payload deployment, which will marginally increase satellite life, but will also increase launch vehicle costs. Similar conflicting technological challenges affect the degree of automation in production, in that automation may decrease production costs, but may also limit future product evolution and customization, common in the industry.

A UNIQUE MANUFACTURING ENVIRONMENT

There are several factors which, when combined, form a very unusual environment with regard to the production of space transportation systems. Most prominent among these factors is the very low production lot sizes and rates that are necessary. Production and launch rates of even 12 vehicles per year are proving to be both challenging to attain and difficult to support, given a very competitive global market for space launch services. Currently the two largest commercial ELV manufacturers are Arianespace , a European consortium, which has a scheduled production run of 70 Ariane 4 vehicles through 1999 and General Dynamics which is working toward the production of 60 Atlas vehicles through 1997. Additionally, Martin Marietta has commitments for over 50 Titan IV vehicles, however these are being produced through a conventional DoD procurement, rather than as a commercial venture. Martin Marietta, as well as McDonnell Douglas with the Delta vehicle, have modest commercial programs which augment DoD contracts in this area.

Related to this factor of small lot sizes is the time spans that are the norm of operation in this industry. In a high technology field "an understanding of the life cycle concept is essential to match product and process life cycles," (Mayer and Moore, 1983, p. 45) however, this assumes that one can define the above. The Atlas, as an example, can be traced back 30 years to its genesis. Current production is scheduled through 1997, but design peculiarities in it could easily have the Atlas flying for another 30 years. Adding to this uncertainty is the fact that new product introduction times are on the order of 3 to 4 years, both due to the engineering tasks required and the sheer magnitude of changing a vehicle, and its corresponding launch and processing facilities, that is hundreds of feet tall. When combined with low launch rates, a product change is a decision which has implications that extend for the better part of a decade, assuming the vehicle is still flying. Clearly this case takes to the extreme the assertion that manufacturing strategy decisions are not only competitively critical but are so structural in nature as to determine corporate success (Hill, 1989, p. 18).

A second factor of distinction is the very high quality sensitivity of the industry. On the commercial side, launch insurance for a single flight ranges from 10-15% of the actual launch cost, with interest penalties imposed on vehicles that have failures and discounts for the more successful. Such percentage fluctuations can have profitable or devastating consequences for an overall business strategy. While DoD procurements are less sensitive to such market forces, incentives and penalties within contracts can be stringent enough that the loss of 2 vehicles out of a 20 vehicle procurement will result in the loss of all earnings for a program. These factors are in addition to the conventional quality issues of customer confidence and economy of material usage.

The final factor distinguishing the ELV industry is the fact that it is, in actuality, a service industry that encompasses a substantial infrastructure. What is contracted for is the delivery of a satellite to a specified orbit. In the past contractors built hardware for either NASA or the Air Force and those government agencies were responsible for the actual launch of the payload, regardless of whether it was a commercial satellite or government owned. The new situation now emphasizes the interrelationship between the launch vehicle and the launch infrastructure. Actual launch operations costs range from 10-20% of the total service cost. The implications for the manufacturing sector are that vehicle design changes to improve manufacturability will likely impact the facilities and equipment used to process and fly the vehicle. This tends to blur the traditional definition of manufacturing strategy as task definition set toward a broader goal (Skinner, 1978). Additionally the service orientation of the industry amplifies the synergy between product availability, market requirements, and satellite weight. For various reasons there is pressure on the satellite manufacturer to increase weight on orbit. This weight is limited, however, by the availability of launch vehicles to deploy a given weight. The result is that satellite weight requirements are driving the growth of launch vehicle design, and , conversely, improvements in vehicle performance cause growth in satellites.

A CASE EXAMPLE: GENERAL DYNAMICS SPACE SYSTEMS DIVISION

In its effort to address the space transportation market, General Dynamics established Space Systems Division in 1987. The primary production programs are the Atlas family of launch vehicles, comprised of the Atlas first stage and a Centaur upper stage, and the Titan/Centaur high energy upper stage, used atop the U. S. Air Force Titan IV vehicle. Shortly after formation of the division a commitment was made to build and fly 18 Atlas I's and sell them commercially. Such a commitment had never been undertaken in the ELV industry without having firm orders before hand, but this was necessary in order to have a production volume of sufficient size to be competitively priced. In 1988 a large commercial order of 11 flights, procured by the Air Force, to launch an entire constellation of satellites enabled the division to expand that production lot to a total of 62. While most of the production of the Atlas was done at the division's San Diego facilities in the past, the commercial environment forced a decentralization of production in order to reduce manufacturing costs. Some component assembly was transferred to a corporate facility in Harlingen, Texas and the first stage assembly was shifted to Vandenberg Air Force Base facilities in order to take advantage of lower labor rates at both sites. Additionally, significant investment was made in upgrading of tooling and a modernization of the vehicle. Due to the climate of perennial production close out, discussed earlier, tooling fixtures had not been upgraded with each special order, but rather they had been adapted to "make due" over the years. Similarly, vehicle modernization had been neglected except where it was functionally required. Lastly, a facility modernization program was undertaken, including the construction of an 80,000 square foot 100,000 class clean room for the assembly of Centaur upper stages used on the Atlas launch system and for the Titan/Centaurs. The net result has been a 4 to 1 reduction in the cost per pound of payload delivered to Low Earth Orbit, and making the Atlas a commercially competitive launch system.

Current manufacturing efforts are now focused on further reducing the cost of production while maintaining the vehicles reliability, as well as driving to attain a 12+ per year production and launch rate. While initial efforts at "restarting" the Atlas production line concentrated on rationalizing and cleaning up the production process, the strategy for future manufacturing improvement is concentrated in three specific areas.

The first of these strategies can best be termed as island of innovation. The cost of bringing the vehicle production capability up to 1980's level technology has been so high that it prohibits a leap into 1990's techniques and technologies of manufacturing. As a result, specific innovations have been identified and ranked according to the order of greatest benefit. The figure of merit utilized in this analysis has been the conventional ROI combined with process quality effects. Those innovations with clear economic benefit are to be implemented in the order of which change will improve quality or enhance mission success the most. An example of this discrete improvement is the planned cellularization of the Centaur tank welding process. The improvements in filmless X-ray quality, and reduction in per unit cost, will allow for an X-ray inspection station to be installed at each weld station. By replacing a single, large conventional X-ray inspection station, the production flow will be greatly simplified and cellular tank welding areas will be established. The emphasis on this strategy of islands of innovation is the application, on a case by case basis, of appropriate existing manufacturing technologies.

The second strategy is the longer term research and development of manufacturing technologies peculiar to the launch vehicle industry. Some of these areas include the evaluation of new, cryogenic tolerant materials that offer benefits in producability, large structure forming techniques, and metal joining techniques applicable to cryogenic hydrogen and oxygen.

A last area is that of strategic acquisitions. The high technology nature of the space transportation industry, combined with the global nature of the market make the acquisition of all or part of certain firms of strategic importance to not only manufacturing capability, but also to the business stance of the division in total. As an example, with a significant portion of the satellite market originating from European countries, a position in the EEC will be vital prior to 1992.

A major effect of the last two points of strategy is that Space Systems Division will, necessarily, increase its level of vertical integration through the decade of the 90's.

NATIONAL IMPLICATIONS

While very low volume production issues of an industry as seemingly exotic as space transportation may appear to be only a discrete curiosity, the future of America in space may hinge on how well the lessons of today are learned. By its sheer nature, the production of space systems will not, in our lifetime, approach the volumes seen in conventional markets. A reality of today's interdependent global economy and the tightening of the U. S. budget is that such low volume production must be made affordable. With the current costs for both deploying and producing space systems the U. S. cannot even afford the most modest space initiatives. Certainly new "next generation" systems will be developed that are more cost effective than those we use today, but the burden now is to understand the dynamics as well as possible so as to ensure that we can afford a future in space.

REFERENCES

Ettlie, John E. Taking Charge of Manufacturing. San Francisco: Jossey-Bass, 1988.

Hayes, Robert H., Steven C. Wheelwright and Kim B. Clark. Dynamic Manufacturing: Creating the Learning Organization. New York: The Free Press, 1988.

Hill, Terry. Manufacturing Strategy: Text and Cases. Homewood, Illinois: Irwin, 1989.

Mayer, Robert J. and Jack Moore. "Applying Manufacturing Strategy Concepts to Practice", *Strategic Management of Operations*, Steven Wheelwright, Roger Schroeder and Jack Meridith (eds), Operations Management Association, 1983, pp 45-50.

Miller, Stanley S. and David C. Rogers. Manufacturing Policy: A Casebook of Major Production Problems in Six Selected Industries. Homewood, Illinois: Irwin, 1964.

Skinner, Wickham. Manufacturing in the Business Strategy. New York: John Wiley and Sons, 1978.

7

MANUFACTURING STRATEGY AND PERFORMANCE OF THE NEW VENTURE FIRM

Richard H. Deane, Vidyaranya B. Gargeya, and Patricia P. McDougall

Department of Management
Georgia State University
Atlanta, GA 30303
(404) 651-3400

ABSTRACT

This research addresses new venture firm performance as a function of consistency between manufacturing and business unit strategy. The empirical study involved 217 new venture firms in the communications and computer equipment manufacturing industries. The results of the study indicate that successful firms exhibit a "consistency" between manufacturing and price-driven business unit strategy. This same consistency was not observed between manufacturing and quality-driven business unit strategy. The research suggests that successful new venture managers are able to correctly relate key manufacturing decisions to business unit strategy. Interestingly, the results of this study tend to indicate that both successful and unsuccessful firms have essentially the same perception of the industry. However, unsuccessful firms lack the ability to match key manufacturing decisions and business unit strategy.

INTRODUCTION

Adam & Swamidass (1989) address several "missing themes" in the operations strategy literature, including the following:

1) Manufacturing strategy research is predominantly one of either identifying strategy process variables and their importance to practitioners (Swamidass, 1986; Schroeder, Anderson, & Cleveland, 1986; and Miller & Roth, 1988), conceptualization of variables (Wheelright, 1984), or axiomatic prescriptions based on case studies (Hill, 1985; and Skinner, 1969 & 1974). Studies that investigate the interrelationship among variables or the effect of a subset of variables on performance are glaringly absent. Studies as conducted by Schoeffler, Buzzel, & Heany (1974) in the business strategy area are absent in the manufacturing strategy area. Swamidass & Newell (1987) provide one empirical investigation of relationships among manufacturing strategy variables.
2) Empirical studies of operations strategy based on objective data are very rare. Significant operations strategy insights could be gained from scientifically designed field stuides that are generalizable. The absence of field data creates a problem for empirical studies because prescriptive writings on operations strategy remain untested. Hence, theory building and validation of prespcriptive ideas have been limited due to lack of empirical studies. At present, there is an acute need for conducting empirical studies as building blocks for operations strategy theory development.
3) Quality is given a high level of attention in the traditional operations management literature; however, it fails to receive this high level of attention in the operations strategy literature. The role of quality as a competitive strategy needs to be studied in greater detail with the aid of empirical research.

It is these missing elements in the operations strategy literature to which this study is addressed. Specifically, this research relates the role of business unit/manufacturing strategy compatibility and new venture firm success.

For more than a decade, both practitioners and academicians have recognized the need for a well-formulated and successfully implemented manufacturing strategy for achieving the company's goals (Buffa, 1984). The literature suggests that manufacturing, as a funtional level strategy should support the business level strategy of the firm. Manufacturing must not only be supportive and integrative within the firm's business unit strategy, but must also be compatible with the other functional unit strategies. By defining how the organization plans to compete, the business unit strategy guides the development of manufacturing objectives and strategies. For example, if the business unit strategy is directed towards price leadership in the industry, then the manufacturing strategy should normally support cost reduction (Porter, 1980). If the business unit strategy is directed towards a high level of quality and customer service, the emphasis of operations should include delivery and flexibility themes.

Case reports suggest that compatibility of manufacturing and business unit strategy is important in determining firm success. The 1911 assembly line strategy of Henry Ford's Model T production facility represents an early example of perfect compatibility between manufacturing and business unit strategies. Ford, embarking on a price leadership strategy, formulated a well-orchestrated effort of high volume production of a commodity product using specialized labor on an assembly line driven by just-in-time materials handling. There are also examples that illustrate poor performance resulting from an incompatibility between manufacturing and business unit strategy. The eventual failure of Laker Airways was certainly caused, in part, by an incompatibility between business unit strategy and operations strategy. Laker originally adopted a low fare, deep discount strategy supported by limited in-flight service, full capacity flights (scale economies), and walk on-walk off boarding (no advanced reservations). Unfortunately, Laker later attempted to modify his business unit strategy to attract more full fare business travelers without the necessary modifications in operations. This change of business strategy resulted in an incompatibility between Laker's low cost operations strategy (and image) and the full service strategy needed to accomodate the business traveler. As an example, the business traveler expected to interact with a full service airline reservations unit, not an "order taker" system. Laker did not offer such a full service, comprehensive reservations system. The same problems were evident in the now defunct People's Airlines, i.e., the inability to attract full fare, full service business travelers with a low cost operations strategy. In manufacturing, a lack of consistency is often evidenced by marketing's attempt to promote the "customized" offerings in a firm's product line, while manufacturing is emphasizing the firm's standardized "off-the-shelf" products through assembly line operations, special purpose equipment requiring long setup times, etc.

Important trade-offs must be made among competing manufacturing objectives; it is impossible to excel in all dimensions simultaneously. This concept was introduced by Skinner in terms of "focus" (Skinner, 1974). The focused manufacturing effort must be one that identifies a limited set of manufacturing goals, while at the same time correlates with the primary business unit strategy of the firm. While a failure to develop a manufacturing focus is one type of strategic error, a misdirected focus is yet another another type of strategic error.

Barring a few studies (Swamidass, 1986; Schroeder, Anderson, & Cleveland, 1986; Swamidass & Newell, 1987; and Miller & Roth, 1988), there is insufficient empirical research in the area of operations strategy. Adam, & Swamidass (1989) point out that operations strategy theory development should use empirical research as building blocks. There is an abundance of anecdotal evidence to suggest that a well-formulated and implemented manufacturing strategy should lead to corporate/business success. However, there are few, if any, empirical studies to support this relationship.

The specific focus of this study is on differences in manufacturing strategy exhibited by successful and unsuccessful firms. It is of interest to determine whether firm success is related to how well the firm's strategic manufacturing choices are integrated with the firm's business unit strategy. The elements of business unit strategy considered in this research are price leadership and quality differentiation. A hypothesized set of compatible relationships is provided in Table 1 (Hill, 1989). The null hypothesis is that there is no relationship between performance and consistency between business unit and manufacturing strategy. A firm's business unit strategy and strategic manufacturing choices are consistent if they exhibit the characteristics given in Table 1.

Table 1: Characteristics of Price and Quality-Driven Manufacturing Process

PRODUCT/FIRM/MARKET/ MANUFACTURING CHARACTERISTICS	BUSINESS STRATEGY	
	LOWEST PRICE	SUPERIOR QUALITY
1. Product Type	Commodity	Speciality
2. Product Range	Narrow	Broad
3. Product Development	Maintaining Current Products	Continued New Product Development
4. Manufacturing Processes	Proven Manufacturing Processes	Innovation in Manufacturing Processes
5. Customer Service	Minimal Customer Service	High level of Customer Service
6. Lowest cost per unit	Over-riding Concern	Not an Over-riding Concern
7. Capacity Utilization	High (Little Excess Capacity)	Low (High Excess Capacity)
8. Processes and Technologies	Public Domain	Patents or Other Proprietary Knowledge
9. Profitability/Market Share	Profitability	Market Share
10. Average Customer Order	Large	Small
11. Top Management	Marketing Expertise	Technical Expertise

RESEARCH METHODOLOGY

This research is part of an on-going, comprehensive study of new venture firm strategy and performance (McDougall, 1987). The research data base was compiled from mail questionnaires, validated in a pilot study, sent to new venture firms within 9 SIC codes in the computer and communications equipment manufacturing industries. Consistent with prior research, a firm was considered a "new" venture if it was eight years old or less (Biggadike, 1979; and Miller & Camp, 1985). The data base consists of 217 new venture firms that were newly started entrepreneurial companies (as opposed to "parented" new venture firms, which were not included in the study).

The mailed questionnaire asked the owner, president, CEO, or chairman of the board to:

1) indicate on a seven-point bi-polar scale the emphasis the business unit placed on each strategic variable when compared to other firms in the industry,
2) indicate on a five-point percentile scale the performance of their business unit as compared

to similar businesses in the industry in the last 3 to 5 years on Return on Sales (ROS). This is based on research that shows self-reported scores on firm performance are good measures of success (Dess & Robinson, 1984). Venkatraman & Ramanujam (1986) as well as others have suggested that, in general, ROS serves as a valid and meaningful measure of firm success.

The sample was checked for representativeness by comparing the research data base with the population of all Dun and Bradstreet new ventures in the industries studied, based on the number of employees and the number of employees and geographical location. A chi-square analysis confirmed that the survey respondents did not differ significantly (p = 0.05 level) from non-respondents based on firm size or location. Additionally, a follow up telephone survey of 23 non-respondents was conducted. There were no significant differences in any of the basic characteristics describing non-respondents and the reserach data base. Reliability was examined using a sample survey of second respondents. Responses from a different top executive in 50 randomly selected firms were received; these responses were used to verify reliability.

The manufacturing posture of the firm was measured through responses to 19 key strategic manufacturing questions with regard to the firm, and 7 strategic manufacturing issues in relation to the industry in which the firms are operating. The details of the 19 strategic manufacturing variables relating to the business unit are given in Table 2. Table 3 shows the particulars of the 7 industry variables.

ANALYSIS AND RESULTS

The 217 new venture firms were categorized into "successful", and "unsuccessful", based on the CEOs self-assessment of their respective firms performance (i.e., whether their firm's performance falls in the top 40%, or bottom 40% respectively in their industry on ROS). The middle 20% were excluded from further study following the polar extreme approach (Hair, Anderson, Tatham, & Gradlowsky, 1979). A summary of the firms' performance ratings are provided in Table 4. Firms omitting performance data were also been eliminated from Table 4 and further study.

A Hotelling's T^2 test (Green, 1978) was first conducted to test the significance of the difference between the centroids of the successful and unsuccessful firms based on the manufacturing and business unit strategy variables, as well as perception of industry variables. The F statistics indicate that successful and unsuccessful firms are not significantly different based on these variables. The details of the F statistics are given in Table 5.

As expected, there is no single strategic theme that differentiates successful and unsuccessful new venture firms in the computer and communications equipment manufacturing industry, e.g., a successful firm is just as likely to follow a price leadership strategy as is an unsuccessful firm. In addition, both successful and unsuccessful firms tend to have the same perceptions about the industry in which they are operating.

Using the Spearman Rank Correlation technique (Daniel, 1978), correlations between the "pricing" strategy emphasis of the firm and 19 strategic manufacturing variables were calculated. The same analysis was conducted for the firm's "quality differentiation" emphasis as well. The results are summarized in Tables 6, & 7 respectively.

As can be seen from Table 6, for firms successful on an ROS basis, four strategic manufacturing variables yielded statistically significant relationships with the firms' emphasis on "price" policy as a business unit strategy. The sign of each correlation coefficient indicates a compatibility in the relationship found with what would be expected from the literature. That is, for successful firms, a business unit which places an emphasis on a "lowest price" strategy correlates with an emphasis on (1) commodity products (as opposed to speciality products), (2) low cost production, (3) public domain processes and technologies (as opposed to patents or other proprietary knowledge), and (4) market share growth. For unsuccessful firms, no statistically significant relationships between strategic manufacturing variables and business unit strategy were noted.

Table 2: Manufacturing Strategy Variables

Each of the following items consists of a pair of statements represents the two extremes on different manufacturing strategic variables. Place an "X' at the position on the scale that best describes your busines units' emphasis on each of them as compared to other firms in the industry.

Emphasis On	Neither Extreme Emphasized	Emphasis On
1. Manufacturing Speciality Products	¦ ¦ ¦ ¦ ¦ ¦ ¦ ¦ ¦	Manufacturing Commodity Type Products
2. Providing a Narrow Range of Products	¦ ¦ ¦ ¦ ¦ ¦ ¦ ¦ ¦	Providing a Broad Range of Products
3. Continued New Product Development	¦ ¦ ¦ ¦ ¦ ¦ ¦ ¦ ¦	Maintaining Current Products
4. Reliance on Proven Manufacturing Processes	¦ ¦ ¦ ¦ ¦ ¦ ¦ ¦ ¦	Innovation in Manufacturing Processes
5. Providing Minimal or no Customer Service	¦ ¦ ¦ ¦ ¦ ¦ ¦ ¦ ¦	Providing High Level of Customer Service
6. Lowest Cost per unit not an Over-riding Concern	¦ ¦ ¦ ¦ ¦ ¦ ¦ ¦ ¦	Over-riding Concern for Lowest Lowest Cost per Unit
7. High Capacity Utilization	¦ ¦ ¦ ¦ ¦ ¦ ¦ ¦ ¦	Excess Capacity
8. Public Domain Processes and Technologies	¦ ¦ ¦ ¦ ¦ ¦ ¦ ¦ ¦	Patents or Other Proprietary Knowledge
9. Existing Channels of Distribution	¦ ¦ ¦ ¦ ¦ ¦ ¦ ¦ ¦	Develop New Channels of Distribution
10. Profitability an Over-riding Concern	¦ ¦ ¦ ¦ ¦ ¦ ¦ ¦ ¦	Market Share Growth an Over-riding Concern
11. Customer Order Small	¦ ¦ ¦ ¦ ¦ ¦ ¦ ¦ ¦	Customer Order Large
12. Sell Products to One Segment	¦ ¦ ¦ ¦ ¦ ¦ ¦ ¦ ¦	Sell Products to Numerous Market Segments
13. No Backward Integration Towards Raw Materials	¦ ¦ ¦ ¦ ¦ ¦ ¦ ¦ ¦	Backward Integration Towards Raw Materials
14. No Forward Integration Towards Consumer	¦ ¦ ¦ ¦ ¦ ¦ ¦ ¦ ¦	Forward Integration Towards Consumer
15. Totally Non-Union	¦ ¦ ¦ ¦ ¦ ¦ ¦ ¦ ¦	Heavily Unionized
16. Technical Expertise Dominates Top Management	¦ ¦ ¦ ¦ ¦ ¦ ¦ ¦ ¦	Marketing Expertise Dominates Top Management
17. Sub-contracting or Sourcing of Production	¦ ¦ ¦ ¦ ¦ ¦ ¦ ¦ ¦	Fully Integrated Production
18. Flexible, Short-term Supplier Contracts	¦ ¦ ¦ ¦ ¦ ¦ ¦ ¦ ¦	Long-term Buyer Contracts
19. Small Scale, Steady Growth Objectives	¦ ¦ ¦ ¦ ¦ ¦ ¦ ¦ ¦	Large Scale, Rapid, Immediate Growth Objectives

Table 3: Industry Manufacturing Strategy Variables

The pairs of statements below relate to competitive dynamics within an industry. Considering only the sectors of the industry in which your business unit has chosen to compete, place an "X" at the position on each scale that best describes the condition within your industry sector(s).

Emphasis On	Neither Extreme Emphasized	Emphasis On
1. High Excess Capacity	|_|_|_|_|_|_|_|_|	Minimal or No Excess Capacity
2. Low Profit Margins	|_|_|_|_|_|_|_|_|	High Profit Margins
3. High Economies of Scale	|_|_|_|_|_|_|_|_|	Economies of Scale Not Significant
4. Low Cost for Potential Customers to Switch Products	|_|_|_|_|_|_|_|_|	High Cost for Potential Customers to Switch Products
5. No Price Wars	|_|_|_|_|_|_|_|_|	Severe Price Wars
6. Patents, Licenses or Proprietary Methods of Production or Operation Not Significant	|_|_|_|_|_|_|_|_|	Patents, Licenses or Proprietary Methods of Production or Operation Critical
7. Customers Not Loyal to Particular Products	|_|_|_|_|_|_|_|_|	Customers Very Loyal to Particular Products

Table 4: Summary of Firm Performance

PERFORMANCE MEASURE	SUCCESSFUL FIRMS	UNSUCCESSFUL FIRMS
Return On Sales	92	52

Table 5: Summary of F Statistics Based on Hotelling's T2

VARIABLES	F VALUE	F CRITICAL
Business Unit and Manufacturing Variables	1.2446	1.5580
Industry Variables	1.7402	2.0100
All Variables	1.4324	1.5640

Table 6: Relationships Between Manufacturing Strategy Decision Variables and Price Leadership as a Business Unit Strategy Using ROS as a Performance Measure

MANUFACTURING VARIABLE	SUCCESSFUL FIRMS	UNSUCCESSFUL FIRMS
1. Speciality Products/ Commodity Products	-0.21843 (p = 0.0365)*	-0.09380 (p = 0.5083)
2. Lowest Cost per Unit is not a Concern/Lowest Cost per Unit is a Concern	-0.47987 (p = 0.0001)*	-0.24396 (p = 0.0813)
3. Public Domain Processes and Technologies/Patents or Other Proprietary Knowledge	0.26328 (p = 0.0112)*	0.04325 (p = 0.7608)
4. Profitability an Over-riding Concern/Market Share Growth an Over-riding Concern	-0.28729 (p = 0.0055)*	-0.21017 (p = 0.1348)

* Relationship is statistically significant at 0.05 level.

The lack of statistical relationships for unsuccessful firms is perhaps more noteworthy than the statistical relationships found for successful firms. That is, unsuccessful firms do not exhibit any consistency between manufacturing strategy and "price" as a business unit strategy.

As can be seen from Table 7, for firms successful on an ROS basis, four strategic variables yielded statistically significant relationships with the firms' emphasis on "superior quality" as a business unit strategy. The sign of each correlation coefficient again indicates a compatibility in the relationship found with what would be expected from the literature. That is, for successful firms, a business unit which places an emphasis on a "superior quality" business unit strategy correlates with an emphasis on (1) new product development (as opposed to maintaining current products), (2) innovation in manufacturing process, (3) high customer service, and (4) top management dominated by technical expertise (as opposed to marketing expertise. For unsuccessful firms, no statistically significant relationships between these four strategic manufacturing variables and business unit strategy were found. However, for unsuccessful firms, statistically significant relationships were found for "superior quality" business unit strategy and (1) patent ownership (2) market share (as opposed to profitability), (3) new channels of distribution, (4) numerous market segments and (5) long term buyer contracts. The sign of each correlation coefficient in Table 7 indicates a compatability in the relationship with what would be expected from Table 1 for the first relationship, patent ownership. That is, firms attempting to gain a competitive advantage through quality differentiation were more likely to also be relying on patents or proprietary knowledge in manufacturing. However, for unsuccessful firms, the relationship between "profitability being an over-riding concern" and reliance on a "superior quality" strategy is contrary to what is expected from Table 1. Furthermore, there is no intuitive or empirical support in the literature to justify the relationships between the third, fourth, and

Table 7: Relationships Between Manufacturing Strategy Decision Variables and Quality Differentiation as a Business Unit Strategy Using ROS as a Performance Measure

MANUFACTURING VARIABLE	SUCCESSFUL FIRMS	UNSUCCESSFUL FIRMS
1. Technical/Marketing Expertise dominates Top Management	0.30500 (p = 0.0033)*	0.13434 (p = 0.3424)
2. Proven/Innovative Manufacturing Process	0.24892 (p = 0.0167)*	0.11288 (p = 0.4256)
3. Low/High Customer Service	0.31326 (p = 0.0024)*	0.22238 (p = 0.1131)
4. New Product Development/ Maintain Current Products	0.22972 (p = 0.0267)*	-0.26539 (p = 0.0572)
5. Public Domain Processes and Technologies/Patents or Other Proprietary Knowledge	0.14827 (p = 0.1584)	0.28611 (p = 0.0398)*
6. Existing/New Channels of Distribution	0.01744 (p = 0.8689)	0.30006 (p = 0.0307)*
7. Profitability an Over-riding Concern/Market Share Growth an Over-riding Concern	-0.01121 (p = 0.9155)	0.30557 (p = 0.0276)*
8. One/Numerous Market Segments	0.00650 (p = 0.9509)	0.29079 (p = 0.0365)*
9. Short/Long Term Contract	0.01260 (p = 0.9056)	0.33104 (p = 0.0165)*

* Relationship is statistically significant at 0.05 level.

fifth manufacturing variables and business unit strategy. This again suggests that unsuccessful firms may have a difficult time establishing the role of quality in business unit strategy.

CONCLUSIONS

This research has provided empirical evidence that illustrates the importance of compatibility between manufacturing strategy decisions and the business unit strategy of the new venture firm. Specifically, the results tend to show that successful new venture firms recognize the need to support a "price leadership" business unit strategy with compatible strategic decisions in manufacturing. However, the results obtained for a "quality differentiation" business unit strategy are less clear compared to those obtained for a "price leadership" strategy. This result suggests that unsuccessful new venture firms, and even to some extent, successful new venture firms, have a particularly difficult time in defining quality differentiation as a business unit strategy, and in relating this strategy to manufacturing.

The results indicate that unsuccessful firms are not able to correctly relate a "price leadership" business unit strategy with manufacturing decisions. For example, unsuccessful firms employing a "price leadership" strategy were just as likely to manufacture speciality products as commodity type products. However, successful firms employing the same strategy were more likely to concentrate on commodity products.

Interestingly, the results of this research tend to indicate that both successful and unsuccessful firms evaluate the industry in approximately the same manner. Successful firms are however better able to make manufacturing decisions that are properly linked to business unit strategy while unsuccessful firms are not.

The reasons for the lack of "focus" of unsuccessful firms can possibly be explained by a combination of the following:

1) Lack of basic understanding of the importance of the relationship between manufacturing strategy decisions and business unit strategy. Such a problem is likely to be especially important for the new venture firm.
2) Inability or unwillingness to "focus" manufacturing resources on a limited set of manufacturing tasks necessary to support price leadership. That is, even though the firm may understand the need to co-ordinate manufacturing decisions, other organizational or internal factors for start up firms prevent a concentration of effort. Such an explanation is particularly plausible for a new venture firm that cannot mount a sustained effort to build economies in the absence of sales volume.
3) The inherent difficulties faced by new venture firms in entering a market based on price leadership. That is, the new venture firm may understand the types of manufacturing decisions that need to be made, but be prevented from executing these decisions because of the intensity of reaction by existing competitors. New venture firms originated by an existing company (not considered in this research) may face fewer difficulties in this regard.
4) The inherent difficulties in successfully defining quality differentiation as a business unit strategy, or the inability of the top management to focus on quality differentiation as a dominanat business unit strategy. That new venture managers apparently declare quality differentiation as a business unit strategy, but continue to pursue cost leadership with certain manufacturing decisions.

Any of the above problems will result in a lack of compatibility between manufacturing decisions and business unit strategy. Additional research is planned to examine, more thoroughly, the relationship between alternate business unit strategies and an expanded set of manufacturing strategy variables, and the differences exhibited by parented new ventures and independent new ventures.

REFERENCES

Adam, E.E., Jr., & Swamidass, P.M. 1989. Assessing Operations Management from a Strategic Perspective. Journal of Management, 15(2):181-203.

Biggadike, E.R. 1979. The Risky Business of Diversification. Harvard Business Review. 57:103-111.

Buffa, E.S. 1984. Making American Manufacturing Competitive. California Management Review, 26(3):29-46.

Daniel, W.W. 1978. Applied Nonparametric Statistics. Boston:Houghton Mifflin Company.

Dess, G.G., & Robinson, R.B., Jr. 1984. Measuring Organizational Performance in the Absence of Objective Measures:The Case of the Privately-Held Firms and Conglomerate Business Unit. Strategic Management Journal, 5:265-273.

Green, P.E. 1978. Analyzing Multivariate Data. With Contributions from J.J. Douglass Carroll. Hinsdale, Illinois:Dryden Press.

Hair. J., Anderson, R., Tatham, R., & Gradlowsky, R. 1979. Multivariate Data Analysis. Tulsa:Petroleum Publishing Company.

Hill, T. 1985. Manufacturing Strategy:The Strategic Management of the Manufacturing Function. London:Macmillan and Company.

Hill, T. 1989. Manufacturing Strategy:Text and Cases. Homewood, Illinois:Richard D. Irwin, Inc.

McDougall, P.P. 1987. An Analysis of Strategy, Entry Barriers, and Origin as Entry Factors Explaining New Venture Performance. Unpublished Doctoral Dissertation, University of South Carolina.

Miller, A., & Camp, B. 1985. Exploring Determinants of Success in Corporate Ventures. Journal of Business Venturing, 1:87-105.

Miller, P.K., & Roth, A.V., 1988. Manufacturing Strategies. Executive Summary of the 1988 Manufacturing Futures Survey. (Manufacturing Roundtable Research Report). Boston:Boston University.

Porter, M.E. 1980. Competitive Strategy:Techniques for Analyzing Industries and Competitors. New York:The Free Press.

Schroeder, R.G., Anderson, J.C., & Cleveland, G. 1986. The Content of Manufacturing Strategy:An Empirical Study. Journal of Operations Management, 6(3):405-416.

Schoeffler, S., Buzzel, R.D., & Heany, D.F. 1974. Impact of Strategic Planning in Profit Performance. Harvard Business Review, 52(2):137-145.

Skinner, W. 1969. Manufacturing ---Missing Link in Corporate Strategy. Harvard Business Review, 47(3):136-145.

Skinner, W. 1974. The Focused Factory. Harvard Business Review, 52(3):113-121.

Swamidass, P.M. 1986. Manufacturing Strategy:Its Assessment and Practice. Journal of Operations Management, 6(3):471-484.

Swamidass, P.M., & Newell, W.T. 1987. Manufacturing Strategy, Environmental Strategy, and Performance:A Path Analytic Model. Management Science, 33(4):509-524.

Venkatraman, N., & Ramanujam, V. (1986). On the Measurement of Business Performance in Strategy Research:A Comparison of Approaches. Academy of Management Review, 11(4):801-814.

Wheelwright, S.C. 1984. Manufacturing Strategy:Defining the Missing Link. Strategic Management Journal, 5:77-91.

8

BRIDGING THE STRATEGIC OUTCOME MEASUREMENT GAP IN MANUFACTURING ORGANIZATIONS

Patricia L. Nemetz

School of Business Administration, MS-46
Eastern Washington University
Cheney, WA 99004
(509) 359-6920

ABSTRACT

Research in manufacturing strategy is necessary to build knowledge about how to compete in a global economy. Many new techniques and technologies are being touted in the literature as cures for industrial performance inertia. However, these techniques and technologies cannot be properly evaluated because of strategic performance measurement problems. This article describes problems with measuring strategic outcomes. A method is then presented for conducting research that requires strategic outcome measurements. One of the major themes is that this method should be temporary; it is far more important for university-industry ties to become closer, and for efforts to be directed at finding practical, standardized methods for measuring strategic outcomes.

INTRODUCTION

Depsite the importance of manufacturing to a nation's economy, there seemed to be little recognition of its value at the organization level prior to the 1980's, where a state of disarray prevailed in many companies' operations. In a landmark article entitled "Managing Our Way to Economic Decline," Hayes and Abernathy (1980), attributed much of the decline in manufacturing to "modern management principles." According to them, the environment of the 1950's had changed from one of relative certainty to one of complexity and instability, but managers did not respond to that change with new practices. The academic community was equally unresponsive to understanding manufacturing needs. "Japanese" manufacturing techniques and flexible manufacturing automation were beginning to change the rules of competition in a way that was very evident to managers confronted with their own companies' performance shortcomings. Meanwhile, academic researchers tended to extend "traditional management principles" by concentrating their efforts on analytical techniques and mathematical modelling of ideal situations. In a recent symposium of academic researchers, it was acknowledged that "what little information exists about the management of factory automation turns out to be interested opinions about how management of automated factories will be different from today. Solid research is needed" (Meredith, et. al, 1986). Equally as important is research into various manufacturing techniques emphasized abroad. According to the symposium researchers, the strategic aspect of operations management may currently be the most serious issue confronting academicians.

Though it has finally been acknowledged that research in manufacturing strategy is sorely needed, a number of difficulties remain in conducting solid empirically-based research. The first of these problems is that business schools and manufacturing organizations have failed to develop close cooperative relationships. This problem is currently being addressed by the establishment of increasing numbers of university-sponsored technology centers throughout the country. Unfortunately, many are concerned only with technology transfer between academic engineering departments and product design groups in industry. Yet, it seems to be managerial practices and manufacturing processes that are in greatest need of improvement. Even when organizations do participate in manufacturing research, they are understandably reluctant to part with proprietary data. Results of research are long term, and it is difficult to pay the price of giving away valuable information when the rewards for doing so are so intangible. The second problem with conducting empirically-based manufacturing research is that the entire field is sorely lacking in measurement standards. The manufacturing environment has changed in such a way that old performance measures are no longer meaningful. Neither the academic nor the industrial community has yet established new standards for assessing general performance or the ability to meet strategic objectives. Several industry-academic cooperative groups, such as CAM-I (Pryor, 1988), are beginning to respond to the need for better performance measurements, but it may be many years before standards are defined, let alone widely implemented. Without publicly reported, standardized measures of performance, there is no straightforward method for conducting manufacturing research.

It is the purpose of this article to provide one method for bridging the gap created by a lack of standards. It is acknowledged that the use of meaningful manufacturing performance standards is a far more objective way to conduct research, but until those standards are created, researchers may have to rely on less objective measures. The paper will address these issues in more detail by first reviewing literature on manufacturing strategy and measurement difficulties. Then an empirically-derived measure of strategic performance will be presented.

MANUFACTURING STRATEGY

Wickham Skinner first coined the term "operations strategy" in 1969 after reviewing the results of several case studies of manufacturing firms (Skinner, 1969). According to Skinner, manufacturing managers lack general management interests and broad, strategic viewpoints, so they tend to emphasize low cost and efficiency in all competitive situations, to the detriment of the company (Skinner, 1969). Though Skinner's calls for action were met with relatively slow development of the manufacturing (operations) strategy concept, some definition was eventually brought to the field both theoretically and empirically. Swamidass (1986) defines manufacturing strategy as the "deployment and development of manufacturing capabilities in total alignment with the firm's goals and strategies. It gives competitive advantage through the best and conscious use of the uniqueness of the manufacturing function" (pg. 472). At an aggregate level, the <u>content</u> of manufacturing strategy is defined by four competitive factors -- cost, quality, flexibility, and dependability (Skinner, 1974, 1978; Hayes and Schmenner, 1978; Buffa, 1984; Wheelwright, 1984).

The first empirical studies of manufacturing strategy were conducted to define content and to test theories about the link between manufacturing and corporate or business strategy. Swamidass (1986) confirmed Skinner's theory that manufacturing strategy is often nonexistent or misunderstood, that corporate and manufacturing strategies are often misaligned, and that manufacturing managers emphasize cost and delivery criteria as most important, while CEO's emphasize quality. Schroeder, et. al (1986), on the other hand, found some relationship between business and manufacturing strategy, although confusion also existed about the meaning of manufacturing strategy. The four most commonly mentioned objectives of manufacturing strategy corresponded to the manufacturing strategy content defined theoretically in the field -- cost, quality, delivery, and flexibility -- although secondary-level objectives were also mentioned. Both studies were a first step in verifying descriptive elements of manufacturing strategy, but each had some methodological problems and called for standardized measures of manufacturing strategy.

As is evident from the above descriptions, major research on manufacturing strategy centered on definitions and explicitness of the concept. Researchers were interested in "intended" manufacturing strategy. A related, and equally important concept is "realized" manufacturing strategy. Realized manufacturing strategy is a measure of a firm's ability to achieve its intended manufacturing strategy. Evidence from Schroeder, et al. (1986) and Swamidass (1986) suggests that intended manufacturing strategy is often ill-defined or nonexistent in many firms, so realized manufacturing strategy may reduce to being a performance measure of undefined but desirable strategic outcomes in many organizations. Over time the concept of intended strategy may become an important starting point in assessing strategic performance but until awareness of the concept becomes widespread among practitioners, its effect cannot be assessed. Researchers may have to rely on measures of strategic outcomes without taking into account where a manufacturing department's strategic emphasis is explicitly placed.

An assessment of strategic outcomes is important because much of the theory around manufacturing strategy now centers on how environmental and technological changes are impacting American competitive capabilities. Skinner (1969) put in motion a manufacturing strategy agenda that emphasized the importance of trade-offs among the strategic objectives of cost, quality, flexibility and dependability. He advised manufacturers to make use of focused factories (Skinner, 1974). Abernathy and Wayne (1974) defined "strategic choice" as "balancing the hoped-for advantages from varying degrees of cost reduction against a consequent loss in flexibility and ability to innovate". Much of the manufacturing strategy field extended the idea of trade-offs by defining product vs. process focus (Hayes and Schmenner, 1978) and process life cycle theory (Hayes and Wheelwright, 1979a, 1979b). By 1981, however, the impact of "Japanese" manufacturing techniques and flexible manufacturing technology were bringing the notion of trade-offs into question. Wheelwright (1981) illustrated the fundamental difference in how the Japanese view the competitive factors of manufacturing strategy and how U. S. theorists view them (Figure 1). The Japanese discard the notion of trade-offs and use techniques that enable multiple manufacturing objectives to be achieved simultaneously (Schlie and Goldhar, 1989). In addition, flexible manufacturing technology has the capability of suppressing trade-offs and allowing competition to be based on simultaneous multiple objectives (Jelinek and Goldhar), 1983). Although descriptions of Japanese techniques and flexible manufacturing technology are beyond the scope of this paper,

Figure 1

U. S. APPROACH TO MANUFACTURING STRATEGY VERSUS JAPANESE APPROACH TO MANUFACTURING STRATEGY

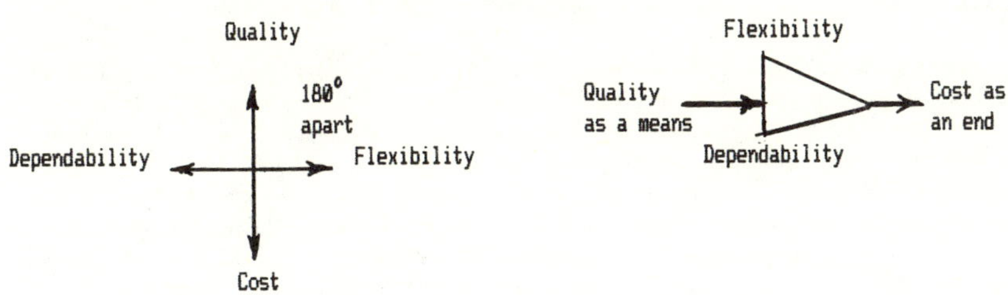

Reprinted from "Wheelwright, Steven, D. (1981) "Japan -- Where Operations Really Are Strategic" <u>Harvard Business Review</u>, Volume 59, Number 4.

Original Source: David E. Kinney, General Electric Co.

they are important developments that beg the question: Is it possible for manufacturing organizations to avoid making strategic trade-offs and simultaneously achieve cost, quality, flexibility, and dependability objectives? To empirically answer this important theoretical question, it is necessary to develop some measure of strategic outcomes.

MEASUREMENT DIFFICULTIES

Developing an instrument to measure strategic outcomes poses several difficult theoretical and practical problems. Such an instrument is basically a measure of multiple performance criteria. Performance measures always present controversial issues, including such disputes as the use of short-term vs. long-term indicators (Banks and Wheelwright, 1979), the use of perceptual vs. objective data (Bourgeois, 1985), and the weight given to different performance objectives desired by different constitutencies (Seashore, 1967; Quinn and Rohrbaugh, 1983, Zammuto, 1984). Performance measures in manufacturing organizations are complicated by additional theoretical controversies that are just beginning to materialize as automation increases.

One such controversy is an outgrowth of problems created by using traditional accounting procedures in an automated flexible manufacturing environment, where product cost techniques do not mirror the manufacturing process. Specifically, distortions in product costing occur because accounting procedures recommend distributing overhead costs in proportion to direct labor or output volume of products (Brody, 1985; Pryor, 1988; Brimson, 1988; Hayes et al., 1988; Hendricks, 1989; Johnson,1989). Such allocation practices are appropriate when manufacturing is labor intensive or based on high volume production. But they are not appropriate in flexible or automated environments where the amount of direct labor is very small (approximately 13% of total cost -- Hendricks, 1989), and most overhead costs are unrelated to incurrence of direct labor. In many automated environments, direct labor and overhead are inversely related; that is, overhead costs related to tooling, machine depreciation, power, etc. come from automation that reduces the need for direct labor. Furthermore, overhead rates based on cost pools of automated and nonautomated departments may be too imprecise to allow accurate costing of specific products travelling through just a few operations (Hendricks, 1989). Overhead rates should be based on costs traceable to specific activities and applied in proportion to how they are incurred. Hendricks (1989) recommends using individual process or machine overhead rates, as well as multiple allocation bases to improve traceability and accuracy of product costs. Yet, according to his survey of 168 Fortune 500 companies, over half use traditional overhead rates and application bases (Hendricks, 1989). As a consequence, many reported product costs are inaccurate and distorted.

A second problem is that many performance measures that were appropriate for tradional environments are inappropriate for today's environments. For example, labor utilization, overhead absorption, and variances from standard, which are appropriate measures for high volume production, may motivate managers to produce unnecessarily for inventory in low volume environments. Nevertheless, Hendricks' survey showed that 80% of the respondents employ standard cost systems (Hendricks, 1989). One of the difficulties associated with establishing performance measures in automated environments is that benefits associated with flexible automation are nonquantifiable (Tepsic, 1983; Meredith, et al., 1986; Greenwood, 1988; Johnson, 1989; Hendricks, 1989). Performance measures should reflect the manufacturing processes, promote decisions congruent with long-term profitability, and help to control operations. Such measures would focus on quality, material control, delivery, inventory, machine performance, and flexibility, as well as accurate cost. But results of Hendricks' survey, which are shown in TABLE I, indicate that many companies do not keep such measures (Hendricks, 1989). Acknowledging that many of the needed measures are subjective and nonfinancial, Hendricks (1989) recommends that representatives from engineering, manufacturing, marketing and management accounting work together to formulate and implement them.

TABLE I

MEASURES USED IN EVALUATING OPERATING PERFORMANCE -- RESULTS OF HENDRICKS' FORTUNE 500 SURVEY

PERFORMANCE MEASURES	PERCENTAGE OF RESPONDENTS	PERFORMANCE MEASURES	PERCENTAGE OF RESPONDENTS
Inventory turnover	75	Machine downtime	48
Labor efficiency	73	Throughput rates	46
Physical scrap counts	72	Warranty costs	38
Cost of rework	72	Number of setups	32
Material usage	61	Product development performance	16
Customer delivery performance	55	Manufacturing flexibility	15
Machine utilization percentages	54	Cost of engineering change orders	12

Source: Hendricks, J. A. (1989) "Accounting for Automation", Mechanical Engineering, February, 1989, pg. 69.

Practical difficulties also exist in collecting manufacturing performance data for research purposes. First, many manufacturing establishments are subunits of larger financial units, therefore, performance data is internal and goes unreported in public financial statements and documents. Furthermore, manufacturing establishments are often reluctant to report performance data to outside sources, because of the need for proprietary safekeeping. Even greater difficulty occurs when the performance data must be corrected for industry, as in the case of cross-sectional research, because the correction must occur by gathering accurate internal performance data from all firms providing similar products.

As a consequence of all these difficulties, the researcher is left with the option of either disengaging from manufacturing strategy research or developing a perceptual measure of strategic outcome. The second option may not be so inappropriate since it is becoming apparent that so many strategic benefits are nonquantifiable and must be subjectively assessed (Tepsic, 1983; Meredith, et al., 1986; Greenwood, 1988; Hendricks, 1989). But the development of a perceptual performance measure must be based on methodological rigor, so that the poor measurements that characterize so much of strategy research can be avoided (Venkatraman and Grant, 1986). The next section describes one such effort to derive an empirically sound strategic outcome measure.

METHODOLOGY

Sample

As part of a larger study to assess the nature of flexible manufacturing technology -- structure relationships, it was necessary to develop a measure of strategic outcomes. Much of the literature on manufacturing strategy and technology is presented in generalities, with a great emphasis on likely trends across all industries. Therefore, the sample was chosen with a view toward finding highly generalizable results. Firms were limited to those that manufacture discrete items (no continuous processors), but they were chosen from a variety of industries. Several resources were used to identify potential research sites in two geographic locations -- southeastern Pennsylvania and the state of Washington. These two locations were believed to provide a variety of types of industries and economic conditions. The Ben Franklin Partnership Annual Report (1986), York Chamber of Commerce Membership List (1987), and Advanced Technology in Washington State Directory (1988), are among the publications used to screen potential sites. The field of possible sites was considerably narrowed by choosing only discrete-item manufacturers with 100 or more employees. Of the 47 that were contacted, 30 agreed to participate (16 in Pennsylvania and 14 in Washington), for a response rate of 64%. Most of the establishments are subsidiaries or subunits of larger corporations (many of them Fortune 500), "thus typifying the dominant operational mode in American industry" (Collins, et al., 1988, pg. 522). The primary industries represented by the sample include transportation equipment, office equipment, heavy equipment, instrumentation, consumer goods, and electrical items. The size of the firms was measured by the number of employees, which ranged from 100 to 17,000. Subsequent to data collection, the distribution of various types of technology in use by the firms in the sample was compared to the results of a large survey of Fortune 500 companies (Hendricks, 1989). The two distributions are similar, so it is believed that the subject sample is an adequate representation of American industry.

A pilot study was executed early in 1988 by conducting in-depth interviews with two manufacturing managers. The interviews were several hours long and included a review of the strategic outcome measure developed for the research. As a result of the pilot study, the measure was expanded to include items of relevance suggested by the practitioners. Data for the remainder of the project was collected between 1988 and 1989 using structured questionnaires, open-ended interviews, plant tours, and retrospective questioning based on information in company and local publications (annual reports, employee newsletters, newspapers, etc.). Face-to-face interviews allowed the collection of a richer data base than would a mail survey, and it is believed that the response rate is better than that which would result from a mail survey. (Several respondents showed the researcher files stuffed with unanswered questionnaires received in the mail). Respondents were middle managers or vice presidents in charge of manufacturing, plant operations, CIM, or computer operations. In twenty-four organizations, two respondents agreed to participate, so a subsample of data was available to test interrater agreement. In addition to these respondents, external respondents provided data for validation of the strategic outcome measure.

Strategic Outcome Measure

To insure that the existence of "intended" manufacturing strategy posed no intervening variable problems, respondents were first asked if they had a manufacturing strategy, and if so, how they would describe it. Attempts to determine intended manufacturing strategy for this project produced findings similar to those of other researchers (Swamidass, 1986; Schroeder, et al., 1986); that is, confusion and uncertainty about its existence. Therefore, since it is likely that explicit intended manufacturing strategy is nonexistent or ill-defined in these companies, the concept of realized manufacturing strategy was not based on the ability to achieve intended strategy. The measure developed is actually a measure of the implicit manufacturing strategy realized, or more simply, a performance measure of undefined, but desirable, strategic outcomes. It is acknowledged that the concept of intended strategy can be an important intervening variable, but until awareness of the concept becomes widespread among practitioners, its effect cannot be assessed.

Because of the virtual nonexistence of accurate, standard, and objective performance data, and the practical difficulties associated with attempting to gather it, it was necessary to rely on perceptual performance measures for the research. Efforts were made to validate the performance data by using a multi-item instrument (available on request). administered to two respondents in a subsample of 24 companies. Items were intended to represent measures of cost, quality, product flexibility, volume flexibility, and dependability. Ratings were based on a 10-point Likert-scale anchored by appropriate indicators; some items were reverse-scored. Respondents were asked to base their ratings on their firm's performance relative to other companies in the industry. Several items representing manufacturing flexibility were drawn from an existing measure developed by Swamidass and Newell (1986). The purpose of building on previous manufacturing research was to begin the process of standardizing questionnaires for future studies.

Respondents were asked for objective evidence as a check on their responses. About 75% of the respondents could provide some objective evidence in the form of trade journal comparisons, company product-line brochures, employee policy descriptions, and annual reports to verify their some of their responses. In addition, local publications reporting on company performance and/or problems were used to verify responses and elicit elaboration when inconsistencies existed. Most such inconsistencies were resolved during the course of the interviews.

An abbreviated performance instrument (available on request) was also administered to a subsample of external experts familiar with a firm's products. Items on the abbreviated instrument were omnibus measures of perceived quality, cost, delivery, and product variety. The experts were usually industrial purchasing agents, equipment specification engineers, or commercial users of the products. They were screened for their expertise by a series of questions about their experience, familiarity with other brands, and technical knowledge. For 5 of the companies, the researcher was unable to locate experts on the products, but a large enough subsample resulted to check for agreement with internal respondents. The questionnaire was administered to the external experts approximately 6 months to a year after the internal data collection in order to allow for organizational lag -- that is, to allow time for the effects of internal decisions to reach the external environment.

Analysis

Analysis included principal components analysis with varimax rotation, checks for interrater agreement (correlations), and calculations of internal consistency (Cronbach's coefficient alpha). Content validity was based on definitions provided in earlier empirical studies (Swamidass, 1986; Schroeder et al., 1986) and on information provided by the two manufacturing managers that participated in the pilot study. In addition, means and standard deviations were calculated to check for meaningful patterns.

These techniques are believed to be appropriate checks on validity for correlational research, as recommended by Venkatraman and Grant (1986). The use of interviews and structured questionnaires provided multiple methods of measuring the same construct. Furthermore, the use of multiple raters, both internal and external, is a means of assessing the veracity of individual evaluations. The use of exploratory factor analysis is a means of determining the distinctiveness of each strategic outcome construct. Reliability measured as internal consistency is an indication of the relationship between observed and true scores.

Results

Results of the principle components analysis are shown in TABLE II. A factor loading of at least .50 was the criteria set for including an item in a factor. In general, all but three items loaded as expected. The item, "speed in changing production volumes", which might have been a volume flexibility indicator, loaded with two other items to form a dependability factor. No separate volume flexibility factor emerged. Two items taken from Swamidass and Newell's (1987) manufacturing flexibility measure failed to load on the flexibility factor for this instrument. The two items are "introducing new production processes" and "R & D effort". There is no reason to expect them to load on what is essentially a product flexibility factor. They possibly represent a fifth factor of "process innovation effort," but since the last factor explained only 7% of the variance and had little theoretical development, it was excluded from further analysis. Future research might address this factor by including more items and better theoretical development.

TABLE II

VARIMAX ROTATED FACTOR LOADINGS OF PERFORMANCE VARIABLES

	FACTORS				
	1	2	3	4	5
	Quality	Product Flexibility	Dependability	Cost	Innovation
1. New Products Introduction	-.173	.720	.003	.135	.010
2. Intro'cing New Prod Proc	.065	-.076	-.032	.359	.356
3. Product Varieties	.002	.649	-.065	-.316	.063
4. Product Features	.079	.636	-.026	-.278	.433
5. R & D Effort	.174	.254	-.039	-.090	.659
6. Product Costs	-.264	-.158	.037	.784	-.147
7. Production Efficiency	.001	.037	.139	.861	.005
8. Product Quality	.888	-.091	.257	-.103	.282
9. Meeting Deliv'y Dates	.025	-.138	.970	.018	.156
10. Speed in Prod Changeover	.161	.747	.075	.357	-.169
11. Product Performance	.779	.049	.084	.010	.058
12. Lead Time of Orders	.229	-.014	.888	.061	-.148
13. Spd in Chg'g Prod Vol	.347	.127	.520	.314	-.271
14. Spd in Dev New Products	.080	.846	-.092	-.165	.182
15. Workmanship	.817	.083	.071	.104	-.022
16. Mat'l Quality	.605	-.037	.039	-.213	.050

Evaluation of internal consistency and agreement among raters must be based on several statistical indicators. In addition to reliability and correlation coefficients, means and standard deviations provide important measures of dispersion and relative ratings.

Results of the validity and reliability checks are shown in TABLE III. In general, most of the variables show moderate to very high internal consistency and internal interrater agreement. For flexibility, cost performance, and dependability, the correlations between external and internal evaluations are lower, but significant. Considering the large number of intervening variables that could influence the agreement between insiders and outsiders (persuasive salesmen, reducing profits to cut price, effect of external distribution systems, etc.), significant correlations are important indicators of consistent performance evaluations. To determine if internal evaluations are biased towards more favorable ratings, it is necessary to compare means of the two evaluations.

TABLE III

VALIDITY AND RELIABILITY COEFFICIENTS FOR PERFORMANCE VARIABLES

Variable	Coeff Alpha	Internal Agreement	Internal/External Agreement
Quality	.83	.48	.15
Product Flexibility	.81	.88	.36 ($p < .05$)
Dependability	.79	.67	.41 ($p < .01$)
Cost	.62	.84	.63 ($p < .001$)

The means in Table IV indicate that external evaluators were consistently more favorable in their evaluations than insiders. This provides some evidence that insiders were not biased in their ratings. Furthermore, the pattern of values is consistent between insiders and outsiders, with cost performance receiving the lowest average rating, and quality the highest average rating.

TABLE IV

SUMMARY STATISTICS FOR PERFORMANCE RATINGS

	Internal Evaluations		External Evaluations	
	Mean	S. D.	Mean	S. D.
Prod. Flexibility	6.79	1.49	7.83	2.03
Cost Performance	4.95	2.10	5.66	2.96
Quality	8.47	.91	8.98	1.00
Dependability	5.79	2.20	7.23	2.44

(Anchors were standard-scaled to signify the following: 1 = poor; 10 good)

Quality performance showed rather low internal interrater agreement, and nonsignificant internal/external agreement. At first glance, this might be explained by the fact that quality can be defined in so many different ways and can mean different things to different people (Garvin, 1986). But there is another explanation for the low value. Respondents had very strong opinions about quality, and the phrase "They (we) are noted for making 'Cadillac' models" was voiced quite frequently. In general, quality was rated quite high for nearly all companies, by both internal and external respondents. Many respondents provided evidence of high quality performance by refering to trade journal or business publication comparisons. Furthermore, a large percentage of these companies used statistical quality control techniques (77%), which Garvin (1986) reports as rare in the U. S. The mean internal response for quality was 8.47 on a scale of 1 (low) to 10 (high), with a standard deviation of .908. The standard deviation indicates that response variance was considerably lower than for other performance variables. External evaluators were even more favorable. The low correlations represent a phenomenon known as "attentuation due to restriction of range." The effect of low variability is to reduce correlations (Allen and Yen, 1979). Because the range of responses was so narrow, confidence can be placed in the level of agreement among raters of quality.

Qualitatively, some of the differences between internal and external ratings can be explained by the fact that outsiders are buffered from problems internal to an organization. Most internal respondents were quite willing to acknowledge that their operations experienced performance variance. Furthermore, they felt a certain subtle but ubiquitous pressure to improve where performance was poor. In general, the internal respondents knew the factors on which they were competitive, and the factors on which they needed to work. To a large extent, these were qualitatively confirmed by external evaluators, although they were unaware of the details and degree of variance. The overall conclusion on the validity of the performance measure is that the combined use of multiple internal evaluators, external evaluators, and external publications provides a high degree of confidence in its validity. Validity is important because the ultimate purpose of the instrument is to test theories on performance trade-offs among manufacturing firms using different technology and manufacturing techniques. Subsequent research examines these relationships.

DISCUSSION AND CONCLUSION

It is becoming increasingly clear that U. S. manufacturers can no longer rely on government intervention to provide a benign economic environment in which to compete. Many new developments in manufacturing techniques and technologies are providing an impetus to examine the means by which U. S. organizations must compete. Unfortunately, no definitive studies exist to explicitly state which techniques and technologies are most appropriate for specific firms. Part of the reason for this paucity of information is that it is extremely to difficult to conduct research when there are no standardized consistent measures of strategic outcomes.

This article provided one method for overcoming the lack of objective peformance measures in manufacturing strategy research. Although the measure developed is perceptual, it is reinforced by a myriad of objective data provided in interviews and by the relatively consistent perceptions of multiple internal and external evaluators. Clearly, single-item objective measures would be far more desirable for research purposes, but the entire manufacturing field is characterized by the use of inappropriate and problematic performance data. Therefore, the tenacious researcher must rely on other methods of measuring strategic outcome performance.

In general, the performance measure developed showed that the largest performance differentiatiors in the sample of firms examined were cost and dependability. Quality and flexibility were generally high. Only 3 firms characterized their production rates as high volume, low mix, whereas 16 reported high volume, high mix, and 11 reported low volume, high mix. No firms reported low volume, low mix. Among firms that indicated performance problems, many reported the syndrome described by Skinner (1969). That is, they were responding to the market demands for product variety, but were unable to maintain an efficient, well-managed factory while doing so. Almost all of the firms received high ratings for quality, from both internal and external respondents. This is quite inconsistent with the findings of Garvin (1986), who reported that quality is considerably lower among American firms than among Japanese firms. Perhaps this is due to the fact that Garvin's (1986) study was limited to a single industry, but more likely is the fact that competition based on quality can occur in many different ways. Garvin (1984) provides at least eight different definitions for competing on quality. Clearly, future research on quality performance must be based on finer-grained measures of quality so that the effects of its underlying multi-dimensional character can be assessed. The performance measure developed for this project resulted in a uni-dimensional measure of quality, which was appropriate for use in the larger research project in which it was used, but other research projects may require more detailed measures of quality.

Future research and practice must also be directed to finding more appropriate performance accounting methods and establishing closer university - industry relationships. The curiosity of researchers and consultants has caused many firms to be "over-surveyed." Consequently, many firms are reluctant to answer time-consuming questionnaires and to participate in studies in which they have no tangible benefit. But there is a clear benefit to be derived from developing more accurate accounting procedures. Many practitioners must be made aware of manufacturing improvements and performance measurement problems through the investigations of researchers; likewise, researchers must have access to large numbers of organizations to investigate manufacturing -- performance links. Closer ties are needed.

REFERENCES

Abernathy, William J. and Wayne, Kenneth (1974) Limits of the learning curve. *Harvard Business Review*, 52(5), 109-119.

Advanced Technology in Washington State Directory (1988). Seattle, WA: Commerce Publishing Corp.

Allen, Mary, J. and Yen, Wendy, M. (1979) *Introduction to Measurement Theory*, Monterey, CA: Brooks/Cole Publishing Company.

Banks, Robert, L. and Wheelwright, Steven C. (1979) Operations vs. strategy: trading tomorrow for today. *Harvard Business Review*, 57(3), 112-120.

Ben Franklin Partnership Annual Report (1986), Bethlehem, PA: Lehigh University.

Bourgeois, L. J. (1985) Strategic goals, perceived uncertainty, and economic performance in volatile environments. *Academy of Management Journal*, 28(3), 548-573.

Brimson, James A. (1988) Bringing cost management up to date. *Manufacturing Engineering*, 100(6), 49-51.

Collins, Paul, D., Hage, Jerald, Hull, Frank, M. (1988) Organizational and technological predictors of change in automaticity. *Academy of Management Journal*, 31(3) 512-543.

Garvin, David, A. (1983) Quality on the line. *Harvard Business Review*, 61(5), 65-75.

Garvin, David, A. (1984) What does "product quality" really mean? *Sloan Management Review* Fall, 25-43.

Garvin, David A. (1986) quality problems, policies, and attitudes in the United States and Japan: An exploratory study. *Academy of Management Journal*, 29(4), 653-673.

Greenwood, Nigel, R. (1988) *Implementing Flexible Manufacturing Systems*, NY: Halstead Press.

Hall, R. W. (1983) *Zero Inventories*. Homewood, IL: Dow Jones-Irwin.

Hayes, Robert H. and Abernathy, W. J. (1980) Managing Our Way to Economic Decline. *Harvard Business Review*, 58(1), 67-77.

Hayes, Robert H. and Schmenner, R. W. (1978) How should you organize manufacturing? *Harvard Business Review*, 56(1), 106-118.

Hayes, Robert, H. and Wheelwright, Steven, C. (1979b) The dynamics of process-product life cycles. *Harvard Business Review*, 57(2), 127-136.

Hendricks, James, A. (1989) Accounting for automation. *Mechanical Engineering*, 111(2), 64-69.

Jelinek, Mariann, and Goldhar, Joel, D. (1983) The interface between strategy and manufacturing technology. *The Columbia Journal of World Business*, Spring.

Johnson, H. Thomas (1989) Managing Costs: An outmoded philosophy. Manufacturing Engineering, 102(5), 42-46.

Meredith, Jack, Hyer, Nancy H., Gerwin, Donald, Rosenthal, Stephen R. and Wemmerloev, Urban (1986) Research needs in factory automation. Journal of Operations Management, 6(2) 203-218.

Pryor, Tom (1988) Issues in the evaluation and selection of advanced manufacturing technologies. TIMS/ORSA National Meeting, April 25-27.

Quinn, R. E. and Rohrbaugh, J. (1983) A spatial model of effectiveness criteria: Towards a competing values approach to organizational analysis. Management Science, 29, 363-377.

Schlie, Theodore W. and Goldhar Joel, D. (1989) Product variety and time based manufacturing and business management: Achieving competitive advantage through CIM. Manufacturing Review, 2(1), 32-42.

Schonberger, Richard, J. (1982) Japanese Manufacturing Techniques: Nine Hidden Lessons in Simplicity. London: Free Press.

Schroeder, Roger, G., Anderson, John, C., and Cleveland, G. (1986) The Content of Manufacturing Strategy: An Empirical Study. Journal of Operations Management - Special Combined Issue, 6(4), 405-415.

Seashore, S. E. (1967) A framework for an integrated model of organizational effectiveness. In K. S. Cameron and D. A. Whetten (Eds.) Organizational Effectiveness: A Comparison of Multiple Models. New York: Academic Press, 55-70.

Skinner, Wickham (1969) Manufacturing -- missing link in corporate strategy. Harvard Business Review, 47(3), 136-145.

Skinner, Wickham (1974) The focused factory, Harvard Business Review, 52(3), 113-121.

Skinner, Wickham (1978) Manufacturing in Corporate Strategy. New York: Wiley.

Swamidass, Paul, M. (1986) Manufacturing strategy: its assessment and practice. Journal of Operations Management -- Special Combined Issue, 6(4), 471-484.

Tepsic, Rudy, M. (1983) How to justify your FMS. Manufacturing Engineering, September, 50-52.

Venkatraman, N. and Grant, John, H. (1986) Construct measurement in organizational strategy research: A critique and proposal. Academy of Management Review, 11(1), 71-87.

Wheelwright, Steven, C. (1981) Japan -- Where operations really are strategic. Harvard Business Review, 59(4), 67-74.

Wheelwright, Steven, C. (1984) Manufacturing strategy: Defining the missing link. Strategic Management Journal, 5, 77-91.

York Chamber of Commerce Membership List (1987). York, PA: York Chamber of Commerce.

Zammuto R. F. (1984) A comparison of multiple constituency models of organizational effectiveness. Academy of Management Review, 9(4), 606-616.

9

Learning Effects in Cell Manufacturing

Eric Svaan, Srikant Datar and Sunder Kekre

Graduate School of Industrial Administration
Carnegie Mellon University
Pittsburgh, Pennsylvania 15213
412 - 268 - 3821

Abstract

Manufacturing cells offer the promise of short lead times and lower costs. In order to fulfill that promise, both technical and human factors must be engineered properly. We report on the experience of a large automobile components manufacturing plant in its initial efforts to implement cell manufacturing. We present an alternative to classic volume-based models of manufacturing learning effects. The scrap costs experienced in six manufacturing cells over a complete model year form the test dataset for model evaluation. The new model shows a better fit to the data. Various enhancements of performance measurement and reporting systems for manufacturing in general, and cells in particular, are discussed.

1. Cells in the Competitive Environment.

Manufacturers today are being forced by competitive pressure to improve drastically in quality and responsiveness. Cell manufacturing is a means to this end; it has been advocated (Hyer and Wemmerlov 1984) as a means of minimizing material flow distance and time, Work-In-Process (WIP) inventory and material handling. All of these affect quality and responsiveness directly in the short term, and also represent cost driver savings which should eventually translate into lower manufacturing costs and consequently improved profitability. Long-term viability of the manufacturing enterprise requires that learning be managed effectively. How do cells serve this long-term objective?

We present a modification of the classic learning curve cost model which recognizes distinctive features of manufacturing cell organization. Following a description of our research site, data from six manufacturing cells is used to test the model. We conclude with a discussion of managerial implications of the study.

2. Cells and Learning.

Besides the objective reductions in space and time, manufacturing cells create an environment conducive to accelerated rates of learning. The potential performance enhancement possible through management action is indicated in a recent article by Schmenner (1988), who surveyed manufacturing firms to correlate productivity improvements with a variety of environmental and managerial variables. Factors that correlate with great significance are quality programs and Just-In-Time (JIT) material movements. By contrast, computerized Material Requirements Planning (MRP) systems are weakly associated with productivity gains. We suggest that an underlying causal factor which explains Schmenner's results, as well as the advantages we observe in manufacturing cells, is the knowledge of the production system required by the program. While MRP systems require the capture and maintenance of massive amounts of data (inventory records, orders and master schedules), detailed knowledge of shop processes other than their time requirements is neither required nor supported by most MRP systems. Quality and JIT programs, on the other hand, require improvement of system operating parameters, not merely capture and maintenance of them. Quality improvement programs establish a context for group problem solving, in which diagnosis and correction of defects leads to refinement of system knowledge and the elimination of the probability of errors. JIT, by eliminating buffer inventories, likewise places a premium on knowledge-based control of production processes. The manufacturing cell, which is the simplest possible production unit given the technology of the product, similarly demands control of the production process.

Cells provide a richer context for learning product-related quality features than does traditional, functional manufacturing. All of the operations required to produce a finished product are arranged together, with visual contact between operators at all the work stations. Feedback concerning quality problems is much more rapid in cells (or other JIT operations) than in traditional manufacturing, due to the minimal WIP inventory and small, typically unit, transfer batches. Job rotation enforces an understanding of the entire production process on the part of each worker. The input material quality criteria that she observes as an operator at Station 2 today become the output quality criteria that she enforces as an operator at Station 1 tomorrow. Her visual and tactile experience supply knowledge in much greater bandwidth than could ever be communicated by drawings or specifications. Where job rotation exposes many workers to each job, the total understanding of the job is not limited by the perceptions of any one person. Co-worker knowledge sharing, both informally on the line and in team meetings, should yield faster learning.

We may summarize the learning advantages of cells:

1. Rapid Feedback. The minimal WIP separating sequential operations means that defects can be identified quickly, and defect-causing process parameters corrected immediately.
2. Whole-part Knowledge. Since products are produced in a local area rather than traveling to remote specialized process areas, the entire process is visible to workers. A better understanding of relevant quality parameters is likely to develop.
3. Knowledge Sharing. Proximity of operators of sequential processes facilitates the sharing of knowledge between workers. Especially when physical arrangement of workstations into cells is accompanied by job rotation among the operators, more brainpower is applied to each step in the process.

We can further quantify the last point. If process knowledge is (conservatively) considered to be a unidimensional variable, and each operator Oi in the set of **m** operators **O** = {O1, O2, ..., Om} has her own self-acquired knowledge K(Oi), then the outcome of perfect knowledge sharing is

$$K'(Oi) = \max \{K(O1), ..., K(Om)\}.$$

If process knowledge is of higher order, then the outcome of knowledge sharing will be

$$K'(Oi) >= \max \{K(O1), ..., K(Om)\}.$$

Learning curves. The classic learning curve studies began in airframe manufacturing, with the observation that the labor Y required to build unit X was proportional to the log of X. This equation can be written

$$Y = kX^{-n},$$

but is more commonly written in log-linear form:

$$\log Y = \log k + n \log X$$

where **Y** represents the unit labor cost, **k** is the labor cost of the first unit produced, **X** is cumulative volume, and **n** is the learning curve parameter. This quantification of the adage that "practice makes perfect" has spawned numerous studies (Dutton and Thomas, 1984). Theory to explain the observed variation in learning rates has been in shorter supply. Manual-paced operations generally show faster direct-labor learning rates than machine-paced operations, but this fact does not give much help to a firm operating within a given technological context, with an existing equipment investment which determines labor content. Recently firms have begun to shift the focus of their cost control efforts away from direct labor. There are good reasons for this shift in focus. First, labor's percentage of total costs has steadily declined. For example, at our research site (to be more fully described below) direct labor costs are approximately 10% of the total cost, whereas material accounts for more than 40%. Also, the majority of operations are machine-paced, meaning that the potential for labor cost improvement is small. Quality efforts, time-based competition, and elimination of waste are the new frontiers of opportunity.Therefore it is apparent that a new area of opportunity for learning studies is to be found in costs of quality.

The new learning model. Our new model revises both sides of the classic learning curve model. Consider a cell staffed by **m** operators. If management establishes a policy of job rotation, the conditions are present for an enhanced rate of learning due to teamwork. The knowledge acquired by each operator can be shared with the others, overcoming the loss of sheer volume experienced by a dedicated operator in a single workstation. On the left hand side, the dependent variable is scrap cost S. On the right hand side, we replace **X** by **XE**:

$$\mathbf{XE} = X/m + (m-1)[X/m]^{\emptyset}$$

where

XE represents "effective volume", to be discussed below
m is the number of operators in a cell rotation group
$\emptyset$ is a teamwork parameter.

Thus the model that we test is

$$\log \text{Scrap} = \log k + n \log XE.$$

Some of the properties of the model that emerge immediately are:

1. When $\emptyset = 1$, XE = X and the new learning model degenerates to the classic model.
2. When $\emptyset < 1$, there are dyssynergies of job rotation. The proper managerial policy in such a case is to forbid job rotation, unless considerations of flexibility compensate for the increased costs.

The parameter $\emptyset$. We postulate that the parameter $\emptyset$ is a function of "three Ts": Technology, Turnover, and Teamwork. When observing mature technologies, in which operating procedures have become standardized, we would expect to find lower values of $\emptyset$ than in newer, less perspicuous technological contexts. This is because procedures where processes are mature are likely to be formalized, with fewer opportunities for learning by the operators. A high rate of turnover in a cell will generally mean loss of knowledge from the group and hence the predicted

$\emptyset$ is lower for high-turnover environments. Finally, the level of teamwork (group problem solving, team meetings, social cohesion) is expected to correlate positively with $\emptyset$.

These observations about $\emptyset$ lead to a characterization of the conditions under which we expect our revised scrap cost learning model to find application. First, the product and/or process must be in a state of flux. Industries in which processes are developing rapidly, such as electronics, would fit here. Also fitting are automotive applications and other fashion and style based industries, in which the processes are generally mature but the product line is turned over frequently. Second, there must be a reasonable degree of stability in the workforce, to preserve the accumulated knowledge in the cell. Third, there must be management policies in place to facilitate the sharing of knowledge.

3. Research Site Background.

Cell vs Traditional manufacturing. Traditional manufacturing in our research site, Plant A, can be defined by its layout and material flow patterns. Equipment and processes are arranged generally according to function. Plastic injection molding, painting, plating, metallizing, hot stamping, welding, and assembly operations each occupy a distinct section of the plant. The plant area exceeds one million square feet. Typical material flow distance during the manufacturing process exceeds 5000 feet, with movements between productive departments assigned to the production control organization. The 1000+ member workforce is unionized.

Master production schedules for the plant are based on customer demand forecasts and orders. They are translated into floor production schedules by OPT (trademark of Creative Output, Inc.) software on running on a mainframe computer. Customer demand is variable, and shift layoffs are commonplace.

Synchronous cell manufacturing at Plant A begins with a specific product-related proposal approved by a joint union-management team. Before startup, jobs in the cell are posted for 10 days. Only two job classifications exist in a cell: **operator** and **job setter**. In contrast, the functions performed in the cell would require as many as six to ten classifications in traditional work centers. Achievement of the classification comes at 45 days, and is based on supervisory evaluation. A 90 day lock-in to the cell assignment applies during cell startup. Operators are expected to be multifunctional, rotating to all jobs within the cell. Cell meetings for problem-solving, continuing job evaluations, and a sense of business ownership are also expected. Training and orientation to cell concepts are provided for supervisors and operators during cell startup. Thereafter, training for new operators coming into the cell is expected to be done by senior operators. Most of the cells are devoted to high-volume products with few setups required. Some robotic handling equipment is in use, but most of the machinery is standalone, with substantial manual intervention required both for production and material movement between machines.

The cells may be classified as semi-autonomous work groups, according to the terminology of Goodman, Devadas and Griffith Hughson (1988), in that the duties performed by cell members are strictly circumscribed by union work rules that reserve, for example, interdepartmental material handling and maintenance activities to members of other labor classifications. Cells do not fit the definition of self-designing work teams, because the freedom of cell operators to define their own task and to take on greater responsibility for vertical activities (such as purchasing, scheduling, etc.) is strictly limited by both management policy and the union contract. Design of the cell is the outcome of negotiations between managers and union representatives who do not work in the cell. The union has historically grieved ad hoc changes in work policy, which inhibits management's encouragement of freedom and experimentation in the cells. The policy of job rotation in the cells, and the extent to which cell operators undertake responsibility beyond that exercised by operators in traditional departments in the plant, indicate growth in the autonomy level of cell teams.

Distinctive characteristics of individual cells. Cell characteristics of product and process are given in Table 1. Several observations are in order concerning the cells. The first concerns the operation type. The two cells designated type A are assembly cells, meaning that all components of their products are either purchased parts or are manufactured (i.e., molded and/or metallized) elsewhere in the factory. They therefore are highly dependent upon other departments for component supply. Both cell 08 and cell 01 have painting operations as well as pure assembly (weld, press-fit, stud insertion) operations. Both of the type A cells were in operation for the entire model year. The products that they make are somewhat similar in that the operations (painting and assembly) overlap. Cell 08's product is a higher-value part, with fewer but much more expensive (and fragile) components does cell 01's product; it has much higher scrap rates per unit but fewer workstations than cell 01.

The two type P cells are somewhat less dependent on other operations than are the type A cells. The type P cells contain molding operations in addition to painting and assembly. Some of their parts still are imported from other departments in the factory, however. In type A and type P cells, the dependence on other departments for certain operations is due to indivisibility of complicated processes which cannot be decentralized economically to cells. One of the type A cells is located next to one of the type P cells (cells 08 and 05). The other two cells are isolated from the others. Cell 10 was in operation for the entire model year under consideration. Cell 05 was in operation for only eight months of that year. The products and operations in the two cells are quite similar. The major difference between the two cells lies in the large product variety in cell 05 (18, the largest number produced in any cell). Cell 10 is the simplest of all the cells, with only one product produced by three operators on each of two shifts. It alone of all the cells has no job setter assigned.

The two type S cells are the most similar pair in the data set. They are located next to each other. Cell 16 was the original location for cell manufacturing of this part type, and was in operation for the entire model year. Its success led to replication which created cell 17. Cell 17 was in operation for only four months of the model year. The parts produced in the two cells are virtually identical in appearance, labor content, required material, and processes. All incoming material for the type S cells is raw material. Molding and assembly capabilities make these cells virtually standalone operations. They come closest to the cell ideal of "Total Control, from Pellet to Sell It."[1] In variety, the type S cells occupy a position between the single-product simplicity of 01, 08, and 10 and the large variety of cell 05. Their products taken individually are the simplest of all those made in cells. The number of operators is small, and in fact of the three workstations in each cell two are identical welding/assembly stations.

Job rotation in the cells. Two of the cells, 08 and 10, meet the ideal of daily rotation among workstations. This means that on approximately a weekly basis, each cell operator works to produce each (of 1) product made in the cell. The workforce (at least on the day shift) in both these cells has remained quite stable

---

[1]It is to our deep regret that pledged anonymity of the location and identity of Plant A prevents the attribution of this slogan, coined by a genius who works there.

over the course of the model year. The cases in which rotation does not follow the ideal pattern are interesting. In type S cells rotation is done on a weekly basis. Since part schedules for the plant tend to call for weekly quantities, this assures that each product routinely made in the cell is seen by each operator each week. Because of the product variety in the S cells, daily rotation might lead to a "hit or miss" operator/workstation/part assignment, with destructive forgetting between exposures. The implication of the rotation practices in type S cells is that a three week cycle is tolerable, when the assignment to a process for a week at a time guarantees experience with all products there.

In cell 01 no rotation is practiced. Cross-training has been tried, but has been difficult to maintain in the face of very high levels of personnel turnover. The large number of workstations means that an operator/workstation/part assignment would repeat once each month on average, taking into account the fact that right- and left-handed parts are produced in batches. The adaptation is to fall back to lower levels of cross-training and informal job-switching rather than full rotation.

The same complicating factors appear to underlie the breakdown of rotation in cell 05. There, the variety of products would lead to a less-than-monthly experience with each operator/workstation/part assignment. The abandonment of rotation is adaptive, in that it avoids forgetting, but with a loss of the "whole part" learning experience.

To summarize, we find job rotation practiced in the cells in which it can be done in such a way as to repeat an operator/workstation/part assignment as frequently as every three weeks. Daily rotation is practiced in contexts that can allow repetition of operator/workstation/part assignment on a weekly basis.

4. Data Analysis.

We collected data from Plant 1 on scrap costs and production volume for an entire model year. The data was fit to the model outline in section 2 above. Prior to running the regression analysis the cumulative volume data was transformed by summing total unit production for the entire year and then dividing by the number of months data available, then dividing the cumulative volume for each month by this value. Mathematically:

$$X_{trans} = X_{raw} / X_{std} \text{, where}$$

$$X_{std} = \text{Total year volume / number of months data.}$$

This resulted in a transformed X value that takes on a value that is close to, but is not exactly equal to, the number of months since the beginning of the model year/start of cell production. Since it is obtained by a linear transformation of the raw volume data, it does not affect the value of the coefficients in a standard regression equation. It does mean that the new log-linear model differs from the classic model, since log XE and log X are not collinear. Hereafter X will be written as shorthand for X_{trans}. The justification for selecting monthly data other than for its computational convenience is that both planning and reporting cycles in manufacturing environments are monthly. No finer granularity of data was available to us, and managers are in any case accustomed to thinking in terms of months.

Regression results. Data analysis was performed by regression. Because we were interested in testing the hypothesis that teamwork affects learning rates, we were simultaneously estimating the parameter $\emptyset$ and the coefficients of the regression equation

$$\log Y = \log k + n \log X,$$

substituting XE for X:

$$\mathbf{XE} = X/m + (m-1)[X/m]^{\emptyset}$$

to compare our new model to the classic learning model. The regressions are summarized in Table 2.

Because job rotation is not practiced in cells 01 and 05, the value of m is set to 1 in those cells. Data were aggregated by type, making the assumption that similarity of process leads to similar learning opportunities across cells. The other bold assumption was that the part similarities within cells justifies the use of aggregated scrap costs per unit produced a fair measure of quality performance. This is justified, in our judgment, by the very fact that products are assigned to the cells based on product and process similarities.

Discussion. The models classic learning curves generate are subject to improvement for cell types P and S, for appropriate values of Ø. Improvement in adjusted R^2 is 29% for S and 123% for P. Furthermore, the classic model (Appendix, Phi=1) shows significance at the .05 probability level neither for the learning factor (coefficient of LOGXE) nor for the regression as a whole (the F-Ratio) for type P and S cells. By contrast, the best fit of the new model is significant at the .05 level both for the regression itself and individually for the learning factor (the coefficient of LOGXE). Because perturbations in the value of Ø only degraded the significance of the regression for type A cells, the conclusion is that no teamwork learning effects are discernable there. The managerial interpretation is that when job rotation can be made to work in manufacturing cells, there can be (but need not be) a significant teamwork effect that makes for increasingly effective learning as time goes on. The best-fit regression equation of the new model is "flatter", less negatively sloped, early in the model year. This corresponds to the startup phase of production, when job rotation means that accumulation of a threshold level of useful production knowledge is proceeding slowly. As the model year wears on, the sharing of knowledge results in dramatic scrap cost reduction.

5. Conclusions and Future Research.

We plan to extend our study by gathering data from cells located in other manufacturing plants. Through more extensive observation and data analysis we hope to understand more clearly the managerial policies that lead to a high value for Ø. It is also necessary to understand the learning rates for non-cell departments engaged in similar manufacturing activity, and to verify that cell manufacturing makes good on its promise to facilitate learning.

We have discussed above the effect of variety of processes and products on the decisions taken in some cells (01 and 05) to forego rotation. We suggest that the number of workstations in a cell need not determine the size of a rotation group. A cell with 5 workstations could, for example, organize itself as a dyad and a three-person group, to reap the benefits of teamwork on a smaller scale and preserve the benefits of cross-training and its consequent flexibility without paying the price of forgetting, from infrequent repetition of operator/workstation/part assignments. To a certain extent this process appears to be going on in, e.g., cell 01, where informal job-switching is practiced rather than full rotation. As an experiment, we set the value of m=2 for cells 01 and 05, to capture some of the effects of team spirit and cross-training even in the absence of full rotation and reran the model. The small improvements in adjusted R^2 indicate that we may want to characterize m as a function of something other than job rotation alone.

We are concerned about increasing the explanatory power of our learning models. We believe that noise in the scrap data may have contributed to the attenuated results we report. Preliminary results from a second test site seem to indicate a dramatically reduced noise level in reported scrap data. We believe that the experience of these cells will be a much clearer demonstration of the learning advantages available in cells, and also a fairer test of our learning model.

Managers as well as academics are vitally interested in improving the accuracy of manufacturing performance measurement of all types, and those that apply to cells in particular. Anticipated bottom-line benefits have not always been recognized by manufacturers after cell implementations. It is not clear in many cases whether the costs associated with establishment of the cells have been recaptured via the cost driver savings. This may be due to actual performance shortfalls in the cells, or to distortions in the performance measuring and reporting system (Banker, Datar, Kekre and Mukhopadyay 1989). For example, the cells we have studied show dramatic (order of magnitude) reductions in the required flow distance within the plant. Logically this should translate into lower material handling costs. These lower costs have not yet been recognized for the products manufactured in cells, pending the adoption of activity-based costing of material handling (a full discussion is presented in Connor 1989). Current cost accounting therefore represents a noisy signalling device concerning the effects of cells. A recent study by Bohn (1988) has identified noise as a significant independent variable which can influence learning rates in semiconductor manufacturing, at the plant level. Given the expense of experimenting with work structures, machine layouts and process flows, investments to reduce the noise in the performance signals of the management reporting system may be extremely cost-effective. Besides the improvements in performance by production operators out on the line, the performance of managers who must authorize organizational change may be enhanced by improved information. Such a managerial performance improvement might have a payoff far larger than the incremental improvements achievable on the plant floor, given the existing organization and technology set.

References

Banker, R., Datar, S., Kekre, S., and Mukhopadyay, T. (1989). "Costs of Product and Process Complexity." Working paper, GSIA, Carnegie Mellon University, January 1989.

Belkaoui, A. (1988). The Learning Curve. Westport, CT: Quorum Books.

Bohn, Roger. (1988). "Noise and Learning in Semiconductor Manufacturing." Working paper, Harvard Business School, May 1988.

Connor, E. (1989) "Production Control Study Document." Working paper, GSIA, Carnegie Mellon University, August 1989.

Dutton, J., and Thomas, A. (1984). "Treating Progress Functions as a Managerial Opportunity." Academy of Management Review, 9, 235-247.

Goodman, P., Devadas, R., and Griffith Hughson, T. (1988). "Groups and Productivity: Analyzing the Effectiveness of Self-Managing Teams." In Campbell and Campbell, eds., Productivity in Organizations. San Francisco: Jossey-Bass.

Hyer, N., and Wemmerlov, U. (1984). "Group Technology and Productivity." Harvard Business Review, 62:4 (July-August 1984), 140-149.

Schmenner, R. (1988). "Behind the Productivity Gains in the Factory." Journal of Manufacturing and Operations Management, 1, 323-338.

Table 1

Cell Information

Type	Cell #	# Products	# Operators	# Shifts	Rotations
A	01	1	9	2	NO
A	08	1*	6	3	DAILY
P	10	1	3	2	DAILY
P	05	18	5	3	NO
S	16	9	3	2	WEEKLY
S	17	9	3	2	WEEKLY

*This product has a single final assembly number. In each of the other cells, the products come in right- and left-handed versions, effectively doubling the number of products made there.

Table 2

Summary of Regression Results

Cell Type	Learning Model	Adj. R^2	F-ratio (d.f.)	Prob	Value of $\emptyset$
A	Classical	.426	9.530 (2,21)	0.001	n.a.
A	New	.426	9.530 (2,21)	0.001	1
P	Classical	.153	2.720 (2,17)	0.094	n.a.
P	New	.343	5.954 (2,17)	0.011	5
S	Classical	.234	3.295 (2,13)	0.070	n.a.
S	New	.301	4.236 (2,13)	0.038	6

Note: Results reported for new model are best-fit values of $\emptyset$.

10

A CONFORMANCE - PERFORMANCE MODEL: LINKING QUALITY STRATEGIES TO BUSINESS UNIT'S PERFORMANCE

K.E. Maani and D.G. Sluti
Graduate School of Business and Department of Management Science and Information Systems
University of Auckland, Auckland, New Zealand

Introduction

In the literature of the 1980's, operational variables such as quality, flexibility, inventory, leadtime and set-up times have been elevated to strategic heights by leading manufacturing researchers and managers.

Of the above variables, quality has emerged as one of the key strategic factors for the survival and success of manufacturing organizations in the new era of global competitiveness. Management philosophies, principles and prescriptions abound in management and quality related literature. Yet despite this dramatic attention to quality, to this date, there exists no theory, model or conceptual framework to integrate and synthesize the vast and diverse ideas and learnings. This shortcoming coupled with a lack of rigorous empirical studies, definitional inconsistencies and measurement difficulties, have resulted in ambiguous and somewhat conflicting research findings. As a result, there are few clear guidelines for manufacturing managers and other decision makers to implement quality management strategies with confidence.

One of the key questions of strategic importance and research interest is how and to what degree process quality (conformance) improves manufacturing and business unit's performance.

The present paper addresses this question by developing a causal model which links process quality, defined as conformance, to manufacturing and business unit's performance via intervening variables of quality costs and manufacturing productivity. The model, henceforth called the Conformance-Performance Model, is discussed in the following sections. The model and its associated hypotheses will be tested using original survey data from New Zealand manufacturing firms. Follow-up papers will describe data collection and measurement instruments, testing and analysis methodologies, and discussion of findings.

The Conformance - Performance Model

The causal chain between quality (conformance) and business unit's performance, through intervening variables of quality costs, productivity and manufacturing performance can be shown in the form of a Path diagram as in Figure 1 below.

Figure 1 - A Path Diagram of The Conformance - Performance Model

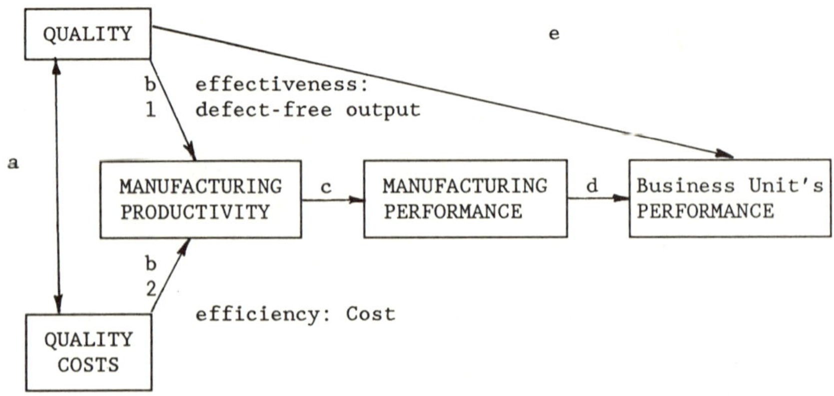

The model variables and their abbreviations are as follows:

Table 1

Quality	(Q)
Quality Costs	(QC)
Productivity	(P)
Manufacturing Performance	(MP)
Business Unit's Performance	(BUP)

The model can be described via six distinct relationships (ie. links) as described below:

Link a: The level of Product Quality (defined as conformance) and quality costs are correlated.

Link b1: Quality and Productivity are directly and positively related.

Link b2: Quality costs have a direct and mixed relationship with Manufacturing Productivity.

Link c: Manufacturing Productivity has a direct and positive impact on Manufacturing Performance.

Link d: Manufacturing Performance directly and positively influence Business Unit's Performance.

Link e: The level of Product Quality has a direct and positive influence on Business Unit's Performance.

The above relationships will be explored in the following sections.

Link a: Conformance and Quality Costs

Although most assertions and studies (see Juran 1980 and Crosby 1979 as examples) suggest that superior quality (conformance) and total quality costs are inversely related, four distinct types of strategies can be envisaged relating level of quality and quality costs. These strategies are illustrated in Table 2.

Table 2 - Conformance and Quality Costs

Strategy	Process Quality	Product Quality	Quality Cost	Main Quality Cost Element	Firm's Expected Position
None	low	low	high	IF, EF	endangered
None	low	low	low	IF, EF	rare
Detection	low	high	high	A, IF	survival
Prevention	high	high	low	P	dominant

* IF = Internal failure (scrap, rework)
EF = External failure (returns, warranty)
A = Appraisal and inspection
P = Preventive

In the first situation, process and product quality are low while quality costs are high. This is due to the lack of a quality strategy resulting in high cost of internal and external failure. As a result, the firm can expect that its productivity, profits, and sales volume will decline. These firms are not likely to survive in a competitive environment.

In certain situations, process and product quality and quality costs may be simultaneously low. This rare situation may occur when both price and customer expectations of quality are very low. An example would be a product with a short life and purchasers who are primarily price sensitive, with little concern for quality. Inexpensive items such as party favors and some disposable materials are examples of these products.

In the third situation, process quality is low while product quality and quality costs are high. This represents the "detection" strategy where main quality costs consist of inspection and internal failure. Here a firm's quality effort is mainly directed towards external quality - to reduce the chance that nonconforming products reach the customer. Inspection and rework rather than "making it right first time" is the main emphasis. While there is concern for quality in such firms, the means of assuring quality with this strategy, is very costly. These firms may survive and even grow where competition is weak. Higher quality of competitive products limit growth prospects for these firms.

In the last situation, the prevention strategy, process and product quality are high while quality costs are low. This occurs when prevention is the main focus of the quality strategy. This is analogous to "do it right first time". Here prevention is viewed as the best way to improve quality while minimizing quality costs. This strategy should enable the organization to compete successfully. As firms move from no quality strategy to a prevention strategy, their prospects for survival and competitiveness will be enhanced.

Manufacturing Productivity

Manufacturing productivity can be thought of having two distinct but related dimensions: (1) effectiveness dimension indicated by volume measures, and (2) efficiency dimension measured by cost (Mali 1978). These two dimensions are investigated separately via causal links b1 and b2 respectively (Figure 1).

Link b1: Conformance and Productivity

The relationship between process quality and productivity is perhaps the most confounded relationship of all. While the traditional view maintains that quality and productivity are conflicting objectives, the emerging view is that the two could be harmonious. It would be premature to discard one statement in favour of the other since evidence and assertions support both views (Maani

1989). Therefore it is prudent to identify the situations and conditions where an inverse or direct link between two variables is likely to be present.

This dichotomous relationship between quality and productivity could be due to the ambiguous definitions of productivity. The traditional term originated as early as 1920's by the U.S. Department of Labor and still dominant in the literature as well as in practice, refers to labor productivity. The emerging view of productivity, however, is much broader and is analogous with waste or loss minimization (Sasaki 1984, Maani 1989).

With some exceptions, more recent studies using the former definitions, report a positive relationship between quality and productivity (Garvin 1983, Wheelwright & Hayes 1985). However, in some situations, mainly characteristic of process industries, an inverse relationship may be observed. For example, in ice cream production one of the dimensions of quality is ice cream density measured by the proportion of ingredients in relation to the amount of air in the ice cream. A good density (conformance) can be achieved by a slow processing (lower productivity), permitting a desirable amount of air into the ice cream. On the other hand, a rapid processing of cream (high productivity) will produce a low density (low quality) ice cream with higher yield. Thus, here, high quality corresponds with lower productivity. This could be viewed as process dominated productivity.

The supporting evidence for a positive correlation between process quality (conformance) and productivity (output rate), however, is more consistent when the broader based definition of quality is used. This harmony is epitomized in the JIT production, where both quality and productivity are seen as one entity: waste-free operations (Hayes 1981, Sasaki 1984, Maani 1989). Further, as Hayes observes "the same conditions which promote defect-free manufacturing operations also increase productivity." That is, a simple product design which is easier to make reduces defects and increases productivity at the same time. Likewise, an improved manufacturing process and technology should result in improved quality and productivity. Also, workers' training and rigorous preventive maintenance are expected to result in simultaneous improvements in quality and productivity.

This view is also propagated by quality gurus, notably Deming (1986) and Feigenbaum (1961, 1983). They maintain that quality improvement is effective in improving productivity because, a quality emphasis often results in increased value of outputs, or increased volume of good output. While, at the same time, the cost or amount of input is decreased.

This favourable relationship has important psychological implications from management's viewpoint as well. That is, it will be easier and more effective to motivate the workforce towards achieving higher quality than to "sell" or impose productivity campaigns (Imai 1986, Sasaki 1984).

Link b2: Quality Costs and Manufacturing Productivity

Here, the second dimension of manufacturing productivity, namely, cost efficiency, is under scrutiny. The causal relationship between quality costs and manufacturing cost efficiency can be seen in the context of quality cost strategy as follows:

a. A quality cost strategy concentrating on preventive efforts positively affects manufacturing productivity. That is, the reduction in input costs realised through a preventive approach, results in higher cost efficiency. A preventive strategy is also less prone to excessive wastes, thereby enhancing manufacturing effectiveness or output productivity.

b. A quality cost strategy focusing primarily upon detection and inspection may have only a neutral or negative effect upon productivity. This approach may result in some increase in good output through rework of identified defectives. However, the cost of input would be expected to rise by a greater amount, resulting in decreased productivity.

   This is characteristic of many "twilight" firms, which although somewhat concerned about quality, they are not prepared to make a total commitment to process and system enhancements which are conducive to structural improvements and lead to strategic advantage.

Link c: Manufacturing Productivity and Manufacturing Performance

This link represents the direct and favorable association between manufacturing productivity and manufacturing performance. This relationship can be analyzed by magnifying various dimensions of productivity and performance. Table 3 shows this relationship.

Table 3 - Relationship between Productivity and Manufacturing Performance

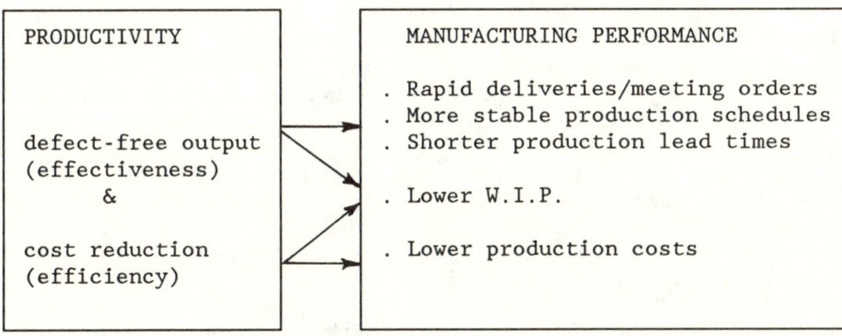

As it is evident from this table, a high output rate combined with lean and defect free operations (cost efficiency) impacts on every dimension of manufacturing performance. More specifically, three relationships can be observed from Table 3.

The first group of performance dimensions correspond primarily with the flow of physical units through the plant. These are directly assisted by an uninterrupted output rate - the effectiveness dimension.

The second category, production costs, are strongly influenced by lean and efficient operations - the efficiency dimension.

The third category, work-in-process inventories (W.I.P.), is affected by both output and cost efficiency dimensions. This can be demonstrated by the following causal chain, originating from quality and preventive cost strategies to the reduction of W.I.Ps.

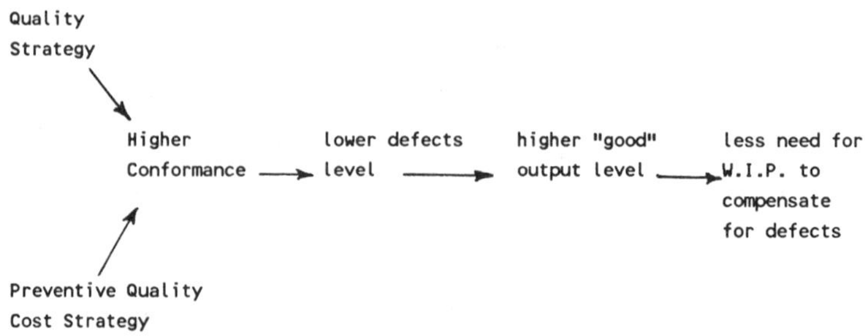

The lower branch of the above chain represents the causal sequence of preventive quality cost strategy which leads to lower failure rates as well as lower total quality costs. This relationship highlights the strategic dimensions of quality and productivity. That is, once quality strategies are translated into manufacturing performance, manufacturing responsiveness and flexibilities are enhanced (ie, shorter cycles). This will enable the organization to change and adapt much more rapidly and to introduce new products much more quickly and effectively. This is where the real strategic benefits are realized.

Link d: Manufacturing and Business Unit's Performance

The benefits stemming from enhanced manufacturing performance translate into superior performance for the business unit and organization. The literature of the 1980's provides ample evidence for this. In fact, the eighties may be remembered as the manufacturing come-back decade. "Competing through manufacturing" has been prescribed by several notable academics (Hayes and Wheelwright 1984 and 1988, Buffa 1984, Skinner 1969) for the ailing western manufacturing organizations.

The dimensions of manufacturing performance outlined in Table 3 have direct and positive impact on firm's prospects to survive and grow in a competitive climate. A lower manufacturing cost position combined with higher quality, greater flexibility and responsiveness affords a firm with a powerful competitive weapon. This strategy has been persistently pursued by Japanese manufacturers in the last four decades with evident success.

The relationship between the operational and the bottom-line performance can be further illustrated by examining the key dimensions of a firm's (business unit or plant level) performance, namely, market share and return on investment. Recent literature (for example see Phillips, et al 1983, Garvin 1983) suggests that the impact of improved manufacturing performance on the bottom-line operates through the following causal chain:

Figure 2 - Relationship Between Manufacturing and Firm's Performance

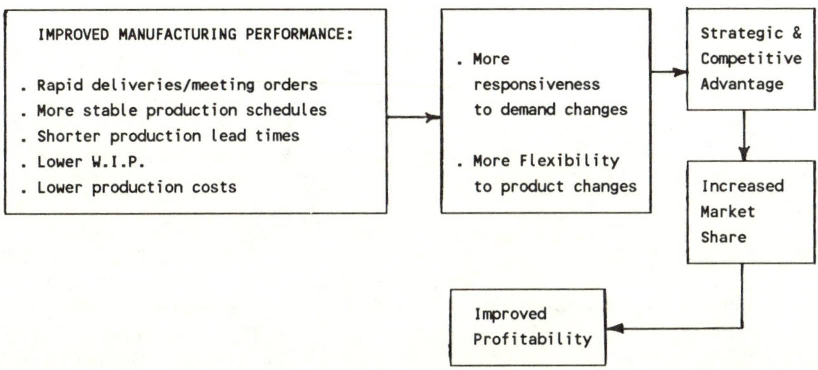

As stated earlier, the link between manufacturing performance and competitive position has been expounded by a number of authors (Hayes and Wheelwright 1984 and 1988, Buffa 1984, Deming 1986, Ishikawa 1985). This relationship holds for both quality and cost-based manufacturing strategies. The causal paths b1 - c - d and b2 - c - d in Figure 1, summarize these strategies respectively.

Finally, the relationship between market share and profitability (ie. ROI) has been empirically investigated by several authors (Phillips, et al 1983, Wagner 1984). These studies consistently confirm the positive association between market share and profitability. In fact, Phillips et al, 1983 conclude that superior quality, through market share as an intervening variable, results in higher profitability (ROI) in all of the six business classifications of the PIMS database.

Link e: Quality and Business Unit's Performance

The link between quality and business unit performance may be explained via two distinct paths, arising from two different definitions of quality:

(1) Manufacturing-based definition or quality of conformance, and

(2) Product-based definition or quality of design.

Manufacturing quality and its link to performance is explained via path (b1 - c - d) in Figure 1. Product-based quality, path (e), characterized by other dimensions of quality such as features, performance or reliability, leads to higher profits through premium prices - a niche marketing strategy.

Conformance quality could also have a direct effect on business unit's performance. That is, a product (and consequently the firm) with a relatively low defect rate, manifested by low customer returns or service calls, develops a reputation for "quality". This perception of "quality" leads to increased sales and enhanced profits. Japanese manufacturing practices and their resulting successes provide ample evidence for this point.

Summary and Conclusion

In this paper, a conceptual model is presented describing the causal relationships between process quality and business unit's performance through intervening variables of quality costs, manufacturing productivity and manufacturing performance. Additionally, a set of hypotheses have been developed for various links in the model. The model provides a framework for integration of diverse ideas and learnings and may be regarded as a step towards a conformance-performance theory.

A distinguishing feature of this research is that it will utilize original data from New Zealand manufacturing organizations. The great majority of previous studies in this area have been based on PIMS database which uses a broad and subjective definition of product quality, often subject to measurement bias. The present study will address this problem by attempting to use objective and consistent measures of process quality instead.

Follow-up papers will describe the data collection and measurement instruments, testing and analysis methodologies and discussion of findings.

REFERENCES

Buffa, E.S. Meeting The Competitive Challenge, Manufacturing Strategy for U.S. Companies. Homewood, Illinios: Richard D. Irwin, Inc. 1984.

Crosby, P.B. Quality is Free. New York: New American Library, 1979.

Deming, W.E. Out of the Crisis. Cambridge Massachusetts: Massachusetts Institute of Technology, Center for Advanced Study, 1986.

Fiegenbaum, A.V. Quality Control. New York: McGraw-Hill Book Company 1961.

Fiegenbaum, A.V. Total Quality Control. 3rd edition. New York: McGraw-Hill Book Company, 1983.

Garvin, D.A. "Quality on the Line". Harvard Business Review, September/October 1983, pp.65-75.

Hayes, R.H. "Why Japanese Factories Work". Harvard Business Review, July/August 1981, pp.57-65.

Hayes, R.H., and Wheelwright S.C., Clark K.B., Dynamic Manufacturing: Creating the Learning Organisation. New York: The Free Press, 1988.

Hayes, R.H. and Wheelwright, S.C., Restoring Our Competitive Edge. New York: John Wiley and Sons, 1984.

Imai, M., Kaizen: The Key to Japan's Competitive Success. New York: Random House, Inc., 1986.

Ishikawa, K.D., and Lu, D.J., What is Total Quality Control? The Japanese Way. Englewood Cliffs, New Jersey: Prentice-Hall, Inc., 1985.

Juran, J.M., and Gryna, F.M. Quality Planning and Analysis, 2nd. edition. New York: McGraw-Hill Book Company, 1980.

Maani, K.E., "Productivity and Profitability Through Quality: Myth and Reality", The International Journal of Quality and Reliability Management, Vol.6, No.3, 1989.

Mali, P. Improving Total Productivity. New York: John Wiley & Sons, 1978.

Phillips, L.W., Chang, D.R., and Buzzell, R.D., "Product Quality, Cost Position, and Business Performance: A Test of Some Key Hypotheses". Journal of Marketing, Vol.46, Spring 1983, pp.26-43.

Sasaki, N. "Rationale for Productivity Increase, Quality Improvement and QC Circles". International Journal of Quality and Reliability Management, Vol.1 No.1, 1984, pp.20-25.

Skinner, W. Manufacturing in the Corporate Strategy. New York: John Lesley & Sons, Wiley, 1978.

Skinner, W. "Manufacturing-Missing Link in Corporate Strategy, Harvard Business Review, May/June, 1969.

Wagner, H.M. "Profit Wonders, Investment Blunders". Harvard Business Review, September/October 1984, pp.121-135.

Wheelwright, S.C., and Hayes, R.H. "Competing Through Manufacturing". Harvard Business Review, January/February 1985, pp.99-109.

11

MANUFACTURING STRATEGY, MANUFACTURING STRENGTH, MANAGERIAL SUCCESS, AND ECONOMIC OUTCOMES

Aleda V. Roth
Duke University
P.O. Box 3018
Durham NC 27713
(919) 684-4188

Jeffrey G. Miller
Boston University
621 Commonwealth Avenue
Boston MA 02215
(617) 353-3458

ABSTRACT

Using a manufacturing strategy framework, this paper has empirically addressed the issue of manufacturing and managerial success and business unit performance. Managerial winners display heightened manufacturing strength on a number of performance criteria simultaneously, including proxy variables for delivery, flexibility, price, and market scope. Operational definitions of manufacturing strategy contents were determined, comprising seven important strategy domains. Five of the manufacturing strategy content areas increase the predicitive power in classifying manufacturing performers over their total strategic capabilities. Our exploratory analysis suggests that manufacturing "superstars," are simultaneoulsy placing more emphasis on advanced manufacturing technology, resource improvements, and quality programs. On the other hand, the current manufacturing "weaklings" are expending more effort on restructuring and integrating information and ystems with most likely the hope of radical improvements in the future.

INTRODUCTION

Over the past two decades, a number of researchers have concluded that manufacturing matters. Manufacturing *matters* to the economy as a whole (Cohen and Zysman 1987); manufacturing *matters* to the business to which it belongs (Abernathy and Hayes 1980; Hayes and Wheelwright 1984; Hayes et al. 1988; Hill 1989; Skinner 1969,1978 and 1985). The belief that manufacturing is an important ingredient in corporate and national success has spurred the development of manufacturing strategies in firms around the world seeking competitive advantage.

However, at this point in time, the linkages between manaufacturing strategy, manufacturing excellence, and economic performance are poorly understood. Research into winning and losing manufacturing strategies is at an early stage. (See DeMeyer and Ferdows 1989, Miller and Roth 1988a, Richardson et al. 1985, Roth1989, Roth et al. 1989, Schmenner 1989, Schroeder et al. 1986, Swamidass and Newell 1987 for recent examples of this type of work, and for reviews of related literature.) Despite logical arguments that manufacturing ought to matter in these linkeages, a clear understanding of how it is related and its relative impact remains to be developed.

The purpose of this paper is twofold. First, it is to propose a general model of the relationships between manufacturing strategy and external economic outcomes such as return on net assets, and profit margin. This is accomplished by using the intermediary performance measures of manufacturing and managerial success to link strategy and economic return. The second purpose of the paper is report on the results of an empirical exploration of several propositions and hypotheses that spring from the general model that is proposed. Data from 193 manufacturing business units from the 1988 Manufacturing Futures Project is used in exploring these propositions.

The General Model

In developing a model of the relationship between manufacturing strategy and economic outcomes, we have taken as given that good manufacturing does not necessarily lead to positive business outcomes, and that positive business outcomes are not always associated with effective manufacturing. It is possible for a firm to be successful with a bad manufacturing strategy, and fail with a good one. In other words, we do not believe that manufacturing matters unconditionally. The general model posited below outlines the general conditions upon which successful economic outcomes are believed to be related to manufacturing.

Figure 1 describes the general model graphically, and relates manufacturing strategy to three different types of performance measures. It also shows the hypothesized relationships between these measures and one another. Empirical studies linking manufacturing strategy and business performance require that operations measures of relative effectiveness be employed as intervening variables (Roth 1989). The first type of performance measure impacted by manufacturing strategy is the relative manufacturing success of the organization. We define relative manufacturing success as the strength of the manufacturing firm compared to its competitors with respect to various manufacturing capabilities such as quality, the ability to compete with low prices, delivery dependability, and flexibility. However, a good manufacturing strategy is not sufficient to guarantee manufacturing success. A firm must also be able to implement that strategy in order to yield relative competitive superiority on key measures. Thus, manufacturing management's abilities, to implement the strategy is included in the model as a moderating variable that shapes relative manufacturing success.

Figure 1. General Model

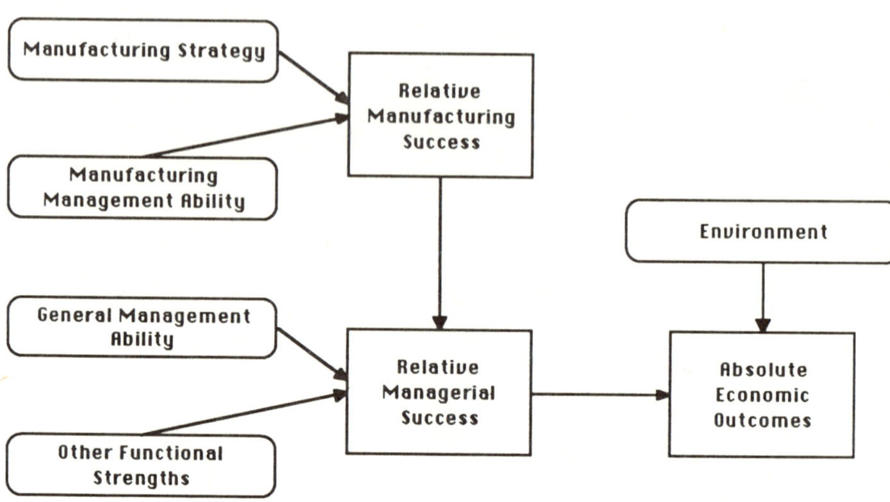

The second type of performance measure indicates the relative success of the management of the firm in utilizing the competencies of manufacturing, as well as of other functions, in acheiving their strategic and business goals. Miller and Hayslip (1989) have commented on the squandered potential that exists in firms that have a well executed manufacturing strategy, but whose general managers are unable to deploy it effectively in conjunction with other functional strategies. Conversely, there are firms which succeed in accomplishing their goals by utilizing strengths in one area (for example marketing or R&D) to compensate for weaknesses in others. The schematic suggests that two key factors in determining managerial success are the relative strength that exists in all areas and that is available to be exploited, and the abilities of the management team to do the exploitation (general management capabiltities).

The final performance measurement type consists of absolute economic outcomes such as ROA and profit margin. These measures are shown to be dependent on relative managerial success and on the environment. An able management team that has and knows how to use functional competencies to advantage, will have good absolute economic outcomes----usually! The qualifier is neccessary because the environment can have a significant impact. A favorable environment can lead to good outcomes despite bad management and poor manufacturing abilities (at least for a while), while a good team with great manufacturing can encounter an unfavorable environment and fail in terms of economic outcomes. For example, chemicals and paper have had a very favorable market environment in recent years in the U.S., while computer manufacturers have faced a more difficult market environment.

Data

The data used in this research is derived from the 1988 Manufacturing Futures Project (MFP) (Miller and Roth 1988b). The annual MFP survey was specifically designed to systematically gather information on competitive factors pertinent to leading manufacturing business units. The underlying logic of the survey instrument was developed at Boston University in 1981 (Miller 1982). Since 1982, the survey has been administered each year in North America, and subsequently, internationally (Miller et. al 1989, De Meyer et. al 1988).

The specific data used in this analysis pertains to the responses of 193 executives from large U.S. manufacturing firms participating in the 1988 study. Details of the conceptual framework and data collection methodology are in (Roth and Miller 1988). Due to the diversity of organizational structures where functional area strategies occur, a manufacturing business unit (MBU) has been selected as the unit of analysis. A MBU is a level in the organization where the manufacturing strategy is integrated with a marketing and product development strategy.

In the 1988 MFP study, 59 percent of the responding executives headed divisions or groups; 26 percent represented entire companies; and the remainder, various other organizational units such as plants. Furthermore, the respondents of the 1988 study represent a broad industrial mix, with 20 percent in the machinery group; 22 percent in electronics; 18 percent in consumer goods; 29 percent in industrial (intermediary materials, parts and components) goods; 11 percent in basic industries.

For the conduct of this study, operational measures of business unit performance (absolute economic outcomes), relative manufacturing success (strength), relative managerial success, and manufacturing strategy contents are required. Since neither business unit, managerial, or manufacturing performance should hinge on a single measure, the analysis becomes considerably more complex. In the analytical framework, it becomes desirable to identify groups of winning business unit performers which have similar performance characteristics on a number of key variables.

ANALYSIS AND RESULTS

Managerial Success

In the questionnaire, we asked respondents to relate their business unit's *average performance* relative to management's goals over the past 2 years on each of the indicators listed below:

- Overall business unit performance
- Overall customer satisfaction
- Profitability (after-tax)
- Return on total assets (ROA)
- Market share of primary product

Since these indicators measure the performance of management against what they reasonably consider can happen, we view them as proxy measures of managerial success.
The average level of business unit performance on each of the indicators was measured on a seven point self-anchoring scale, where "1=substantially worse than goal level performance," "4=goal level performance," and "7=substantially better than goal performance." Since they are relative measures, these qualitative indicators control for the size of the MBU, the industry group, and other important variables which might artificially introduce biases. Their major limitation, of course, is that the relative performance values are based upon the perceptions of the executive completing the survey, and as such, are subject to measurement errors. For this reason, further validation of these metrics against absolute, qunatitative performance is critical.

Table 1. Relative Managerial Success of MBU Performance Groups

QUALITATIVE MBU SUCCESS VARIABLES*	MBU PERFORANCE GROUP CENTROIDS Total Sample N=180	Losers N=78	Winners N=95	In Between N=7	F (Probability)
TAXONS**					
Overall Performance	4.78	3.60	5.73	5.14	122.3 (p<0001)
Overall Customer Satisfaction	4.89	4.44	5.34	4.00	25.2 (p<0001)
Profitability (after tax)	4.56	3.10	5.73	5.00	148.9 (p<0001)
Return on Assets (ROA)	4.53	3.31	5.53	4.57	135.2 (p<0001)
Market Share (primary product)	4.56	4.06	4.93	5.14	21.1 (p<0001)
OTHER MEASURES					
Sales Volume in Units	4.70	3.90	5.33	4.71	36.7 (p<0001)
Sales Revenues	4.63	3.64	5.42	4.86	65.0 (p<0001)

*Average (MBU) performance relative to goals over the past 2 years (1987-88);
(Interval scale (1-7), where 1= substantially worse than goal performance,
4= goal level performance, and 7=subtantially better than goal performance)
**Cluster criterion variables

The five qualitative MBU performance variables served as "taxons," or clustering variables. To classify groups of MBUs based upon similarity over the taxons, a clustering procedure using the agglomerative average linkage algorithm was adopted (Green 1978). The determination of the number of clusters which provide the most interpretable portrayal of the data remains a difficult problem (Milligan 1985) Models deriving from two to four clusters were obtained. Two criteria were employed to determine the final number of clusters to be used in subsequent analysis. First, the number of clusters was between n/30 to n/60, where n is the sample size. Second, we looked for the managerial interpretability of the clusters on the defining variables using ANOVA and the post hoc pairwise comparison tests of mean (centroid) differences.

As a result of this procedure, three managerial performance clusters emerged. Two of these clusters revealed clear cut "managerial winners" and "managerial losers" over each taxon, using the group centroid (mean) as a measure of comparison. (See Table 1.) A small third cluster of "in between" business units also emerged in the data. Because of the relatively high inter-item correlations among the taxons, the formation of two dominant business unit clusters was expected. Thus, the third cluster was considered to be comprised of outliers, and was dropped in subsequent analyses. Table 1 shows that the managerial winners and losers are significantly distinguished on two other nonfinancial performance variables which were not used in formulating the taxonomy. These variables were withheld to aid in validating the construct integrity of the managerial performance clusters. Namely, the winners were much more likely to meet sales volume and revenue goals than the losers.

The general model discussed previously suggests the following hypothesis:

H(1): Management winners will have better absolute economic outcomes, expressed as profits and ROA's, than losers.

In Table 2, the managerial performance groups are characterized by their average values over various economic and other quantitative measures collected in the survey. There is strong statistical evidence to support the hypothesis that the managerial winners' group has a higher ROA than the losers, 23.6% versus 12.8%, respectively. Weaker evidence exists for group differences on the average net pretax profitability of the business unit, 11.4 % for the winners in contrast to 7.8 percent for the losers.

Statistical significance was determined using a two-way fixed effects model ANOVA, using the number of plants in the MBU as a covariate and managerial performance cluster and MBU industry group as main effects. This procedure was adopted to minimize bias that might arise in the absolute measures as a result of the MBU size and the industrial mix of the sample population. Since no two way interactions between managerial performance group and industry were observed, the main effects of the managerial performance clusters and their significance levels are reported separately for each quantitative performance measure.

Table 2. Economic Outcomes and Other Measures by MBU Performance Groups

QUANTITATIVE PERFORMANCE VARIABLES*	Total Sample	MBU PERFORMANCE GROUPS Losers	 Winners		
		(Average Absolute Performance)		F	(Probability)*
Annual Sales Revenue (Millions$)	$711.0	$755.9	$703.4	3.37	(p=.68)
Net Pretax Profits (% of Sales)	9.9	7.8	11.4	3.51	(p=.063)
Research & Development (% of Sales)	4.1	4.3	3.6	0.10	(p=.75)
Unit Growth Rate in Sales (%)	10.0	8.9	10.9	0.31	(p=.58)
Pretax Return on Assets (ROA %)	20.2	12.8	23.6	9.39	(p=.003)
Market Share of Primary Product (%)	37.4	39.4	35.1	0.92	(p=.34)

*Observed F statistic and probability of the MBU performance group main effect in a two-way fixed effect ANOVA design, controlling for size of the MBU. The main effects of each ANOVA are (1) MBU Performers, (Losers/Winners) and (2) Industry Code (Consumer, Industrial Goods, Basic Industries, Machinery and Electronics) with the number of plants in the MBU as a covariate. No interactions were significant. Size was was significant for sales revenue. The industry effect was significant ($p<.05$) on: sales revenue, profits, R&D, and ROA. (See Roth and Miller 1988 for means).

Table 2 also shows the results of similar statistical tests to see if managerial success was related to other presumed determinants of business unit success, such as market share, R&D expenditures, overall sales, and unit growth rate. While absolute economic success may be influenced by these factors, Table 2 shows that they are not significantly associated with managerial success as it is measured here. This suggests that the operational measures of management success employed in this study are independent of other "quantitative" factors that may influence economic outcomes.

Manufacturing Success

To gauge relative manufacturing success, the manufacturing executives were asked to describe the strength of their competitive capabilities relative to their primary competitors in the same industry . Eleven critical capabilities were presented to the manufacturing executives. Using the principal components method of factor analysis with varimax rotation, the eleven original variables were reduced to five independent dimensions of manufacturing strength. Table 3 presents the eleven original variables comprising the five strength dimensions. (See Green 1978 for a technical discussion of the multivariate statistical procedures used.) Each manufacturing strength dimension has face validity, and reasonable internal reliability for newly developing scales. They are consistent with factors tapping manufacturing capabilities found by Huete and Roth (1987).

Table 3. Operational Measures of Manufacturing Success

STRENGTH DIMESIONS AND ORIGINAL VARIABLES**			
	Loadings***		Loadings***
Quality (.70)		**Price**	
Consistent Quality (Reliability)	0.79	Offer Low Prices	
High Performance Products	0.76	**Market Scope (.74)**	
Delivery (.77)		Broad Distribution	0.86
Dependable Delivery Promises	0.88	Advertise Effectively	0.78
Fast Deliveries	0.83	Broad Product Line	0.64
Flexibility (.53)		After Sales Service	0.62
Design Chg /Intro New Products	0.92		
Rapid Volume Changes	0.52		

*(Cronbach Alpha) Inter-item reliability.

**For each original variable, the firm's strength relative to primary competitors. Interval scale (1-7), where 1=substantially weaker and 7=substantially stronger.

***Factor loadings are the correlation of the variable with each strength dimension.

The resultant five strength dimensions are proxy variables corresponding to theoretically important manufacturing performance criteria. They cover the traditional manufacturing performance constructs of "quality," "delivery," "flexibility," and "price" which are commonly touted as primary linking pins between business unit strategy and manufacturing strategy. The fifth dimension corresponds to a boundary spanning capability, which we label "market scope" . The concept of market scope was first observed in Miller and Roth (1989). Market scope represents a set of value added capabilities which transcend the MBU and link functional capabilities to the customer.

Using scales derived from each of the five manufacturing strength dimensions as taxons, the average linkage method of cluster analysis was once again applied to the 193 cases in the 1988 MFP data base. The same criteria for the determination of the number of clusters was adopted as previously cited for choosing clusters of managerially successful firms. Three manufacturing strength clusters were formed, which we labelled: "weaklings," "superstars," and "middlemen," respectively. (See Table 4.) Oneway ANOVAs over each defining taxon show that the weaklings scored worse on every manufacturing performance dimension; the superstars' strength exceeded the weaklings in every case. The middlemen were similar to the winners on quality, flexibility, and market scope but lagged them on delivery and price. A separate Chi Square test of statistical significance rejected the hypothesis that there was an industrial bias associated with the manufacturing strength clusters at the .10 level. Since the manufacturing performance groups are not contaminated by industry differences, this finding adds to the practical utility of the clusters for subsequent analysis.

The general model outlined earlier suggests the following hypothesis regarding the relationship between manufacturing and managerial success:

> H(2): Greater manufacturing strength will be associated with those firms with the highest levels of managerial success.

Table 4. Relative Strategic Strength of Manufacturing Performance Groups

STRATEGIC STRENGTH DIMENSIONS	MANUFACTURING PERFORMANCE GROUP CENTROIDS				
	Total Sample	Weaklings	Superstars	Middlemen	
	(Average Strategic Strength)				
	N=180	N=89	N=78	N=25	F (Probability)
TAXONS*					
Quality	5.39	4.72	6.02	5.78	57.5 (p<0001)
Delivery	4.95	4.44	5.78	4.14	59.0 (p<0001)
Flexibility	4.56	4.00	5.16	4.88	39.2 (p<0001)
Price	4.12	4.11	4.65	2.96	19.1 (p<0001)
Market Scope	4.76	4.13	5.29	5.34	47.7 (p<0001)

*Cluster criteria scale variables. (See Table 3 for defining scale variables.) Interval Scale (1-7), where 1=average strength substantially lower than competitors and 7= strength substantially greater.

The rationale behind this hypothesis is that the manager in a firm with high levels of manufacturing strength will have more options to play with in meeting their goals and objectives. Ceteris paribus, being better at manufacturing improves the odds of being managerially succesful, and that improves the odds of having good economic outcomes.

There are two ways of testing this hypothesis with the data at hand. The first is to employ a oneway ANOVA design. In all, five models were tested where the MBU managerial performance group (winners versus losers) was the independent variable, and where the MBU's score over each of the five manufacturing strength dimension scales comprised a dependent variable. Table 5 highlights the results. On every strength dimension, with the notable exception of quality, there was a statistically significant observed difference in the relative manufacturing capabilities of management winners versus losers.

Table 5. Relative Manufacturing Strength by MBU Performance Groups

STRATEGIC STRENGTH DIMENSIONS	Total Sample N=180	MBU PERFORMANCE GROUPS Losers N=78	Winners N=94	F (Probability)
		*(Average Strategic Strength)**		
Quality	5.39	5.28	5.45	1.21 (p=.270)
Delivery	4.95	4.73	5.19	7.39 (p=.007)
Flexibility	4.56	4.34	4.72	6.08 (p<.015)
Price	4.12	3.91	4.40	6.34 (p<.013)
Market Scope	4.76	4.57	4.99	8.76 (p<.004)

*Average strategic strength on each dimension. (See Tables 3 and 4.) Interval scale (1-7).

The lack of significance between managerial winners and losers on their strength on quality was contrary to our initial expectations. Quality has been a dominant theme in manufacturing over the past decade (Miller and Roth 1988ab). It has held a tenacious grip as a top competitive priority for all manufacturers. However, these findings suggest that "quality" is a prerequisite for entering the competitive arena. While important overall, it's value as a competitive differentiator may be waning.

The dominance of delivery strength corresponds to the increasing importance of time-based competition in manufacturing, including delivery dependability with respect to schedules and delivery speed. Managerial winners not only can deliver on time and with more speed than their competitors, but they also have the delivery system capabilities to offer a broader range of products. They can reach out and touch their customers through distribution and after-sales service. This finding is consistent with that of Schmenner (1988).

Market scope strength gives MBU performers the opportunity to have quickened market feedback on their products. In related research using the 1987 MFP survey, Miller and Roth (1989) found that major differences between categories of strategy are accounted for by market forces. Looking to the future, Chase and Garvin (1989) suggest that the factory of the future will be a market oriented "service factory." As such, a market-driven manufacturing orientation may yield a significant competitive advantage.

The ability to offer low price is a more traditional manufacturing capability. As a traditional measure of manufacturing performance, lowering production costs typically plays a prominent role in manufacturing environments. Yet, while the managerial winners have relatively more strength on this dimension than the losers, low price capabilities are rated the lowest overall. We do not know from these data whether low price capabilities are not perceived as generally important or whether they are most difficult to achieve.

MBU winners also appear to possess a higher degree of flexibility than their counterparts who are not faring as well. Achieving manufacturing strength on flexibility has also emerged over the past decade as a critical need for manufacturing managers. (De Meyer et. al 1988, Gerwin and Tarondeau 1989). It is one of the more difficult capabilities to achieve (Roth et. al 1989).

In sum, U.S. business unit winners appear to be competing on a number of competitive factors simultaneously, including price and flexibility. The study does not provide empirical support for the "trade-off" model among important competitive priorities. The superstars appear to be doing it all. This finding is consistent with the position taken by Edmonson and Wheelwright (1989) in their review of the future competitive demands on manufacturers.

The second way of testing hypothesis 2 is to examine the extent to which classification as managerial winners/losers and superstars/weaklings correspond. Table 6 provides a crosstabulation of the two taxonomies.

Table 6. Number of Managerially Successful Firms by Manufacturing Success Groups

Performance Groups	Weaklings	Superstars	Middlemen
MBU-Losers	48(13.1)*	20(13.8)	10(8.83)
MBU-Winners	31(22.5)	50(26.6)	13(12.4)

*The data in parentheses (...) is the mean ROA associated with each subcategory. $X^2 = 15.6$ ($p<.001$)

Table 6 shows that manufacturing weaklings are much more likely to be managerial losers, and that superstars are much more likely to be winners. These findings further demonstrate that the economic returns are highest for those that score high as manufacturers and managers, and substantially lower for weak manufacturers who also have weak management success. The surprise is that the economic returns of very good managers (*winners*) are also high, even when the organization is weak from a manufacturing point of view. Given a choice between being a great manager and a world class manufacturer, the data suggest the former is preferable.

The weaklings/winners and the superstars/losers are fascinating combinations and deserve much more exploration empirically. How do good managers get the best from their organizations even when manufacturing is not strong? Why can't the "excellent manufacturer's" always translate their business success into economic success?

Manufacturing Strategy Contents

In this research, building upon earlier related empirical research (Roth 1987 and 1989), a manufacturing strategy is defined as "pattern stream" of first order manufacturing decisions. Under the rubric of the Hayes and Wheelwright (1985) schema for manufacturing strategy, the analysis considers the content of a manufacturing strategy in terms of first level structural ("brick and mortar") and infrastructural ("soft") decisions. Over a set of 39 key action programs covering major portions of the structural and infrastructural strategy content domains, the respondents to the 1988 Manufacturing Futures Survey were asked to indicate whether or not each designated program was currently (as of 1988) being given a significant degree of emphasis in order to improve operations effectiveness.

Table 7. Operational Measures of Manufacturing Strategy Contents

MANUFACTURING STRATEGY CONTENT DIMENSIONS AND ORIGINAL VARIABLES**

Structural Contents	Loadings***
Materials Flow-JIT (.77)	
Manufacturing Lead Time Reduction**	0.79
Vendor Lead Time Reduction	0.76
Just-in-Time (JIT)	0.57
Reducing Setup/Changeover Time	0.55
Advanced Process Technology (.60)	
Flexible Manufacturing Systems	0.58
Group Technology	0.57
Robots	0.55
Computer Aided Design (CAD)	0.44
Capacity Upgrade (.49)	
Capacity Expansion	0.62
Reconditioning of Physical Facilities	0.55
Automating Jobs	0.50
Restructuring (.64)	
Reducing Size of Manufacturing Units	0.70
Plant Relocation	0.60
Manufacturing Reorganization	0.58
Closing Plants	0.56
Changing Labor/Management Relationships	0.48

Infrastructural Contents	Loadings***
Resource Improvements (.68)	
Supervisor Training	0.76
Management Training	0.71
Worker Training	0.67
Preventative Maintenance	0.60
Job Enrichment	0.50
Worker Saftey	0.45
Quality Programs (.61)	
SPC-Product	0.75
SPC-Process	0.66
Vendor Quality	0.51
Zero Defects	0.46
Information and Systems (.67)	
Integrating Sys Across Areas	0.74
Integrating Sys Within Mfg	0.74

*(Cronbach Alpha) - Inter-item reliability

**For each original variable, those programs being given a significant degree of emphasis.

***Factor loadings are the correlation of the variable with each dimension.

Of the 39 programs represented in the survey, seven primary dimensions of manufacturing strategy contents were observed over 28 of the variables. (See Table 7.) The final 28 variables included as a measure of a particular type of manufacturing strategy content, had no double loadings on any factor and were consistent with the underlying strategy construct using the measurement approach of Churchill (1979). Strategy content scales were formed by averaging the defining variables of each dimension so that each content area represents the percent of the defining variables which were checked. These strategy contents were almost identical to those found by (Roth 1989 and Ward, Miller and Vollman 1989)

To determine whether particular manufacturing strategies were generically associated with manufacturing strength, the multivariate procedure of stepwise discriminant analysis was performed. Manufacturing strength clusters of "weaklings" and "superstars" served as the dependent variable. Independent (predictor) variables included four of the five industry groups defined in the survey, coded as dummy variables, and scales on each of the seven manufacturing strategy content dimensions. The 5th industry group (the basic industry group) was held out as the industry base. Since manufacturing content variables have the potential for being industry dependent, the inclusion of industry groups into the predictive model is critical.

Six variables were found to have a significant multivariate relationship with manufacturing strength. The industrial goods sector was the only significant industry effect that emerged. In contrast to the basic industry group, membership in the industrial goods sector was likely to correspond to total manufacturing strength. Of the seven manufacturing strategy content variables represented by the scope of the 1988 MFP survey, five were significantly related to building total manufacturing strength: Resource Improvements (.52), Advanced Process Technology (.37), Quality Programs (.48), Restructuring (-.56), and Integrating Information Systems (-.55). (Figures in (..) are the standardized discriminant function coefficients.)

The resource improvement dimension relates to the degree of attention being placed upon human and physical resources of the business unit. Care and nurturing of total manufacturing resources exhibits the strongest positive correlation with overall manufacturing strength. Emphasis on tri-level training (management, supervisors, and workers), as opposed to uni-level or bi-level training, is a strong component of the resource improvement. We speculate that superior manufacturing performance is built upon "the upgrade and maintenance" of the manufacturing knowledge workers - knowledge captured at all levels of the function.

Quality programs are a strong predictor of manufacturing strength. Since we earlier found no difference in the univariate analysis between MBU managerial performance groups on their overall quality strength, could this be further support for our assertion that quality management is a manufacturing basic? These data suggest that that manufacturing superstars are more consistent with respect to their emphasis on formal quality programs.

Manufacturing superstars are placing significantly more effort on advanced manufacturing processes than the weaklings. Having quality under control, and the development of fundamental manufacturing human resources may be seen as the prequalifying factors which must coincide with the successful implementation of advanced manufacturing technologies. Under these conditions, the application of advanced technology is likely to contribute to superstar strength, both now and in the future. Emphasizing advanced technology without the supporting infrastructure may portend a certain disaster, one which the superstars appear to be avoiding.

Two manufacturing strategy contents displayed negative relationships with manufacturing strength. The superstars are currently not emphasizing restructuring nor are they placing weight on integrating information systems. That restructuring is not occurring in the superstar group may be an indication that when structures are working well, manager's don't attempt to change them. The weaklings are placing more emphasis on various means of restructuring their manufacturing units, as they apparently perceive the need for radical surgery.

The inverse relationship between manufacturing strength and the propensity to integrate information systems was not anticipated. Information is one means to control environmental uncertainty. It is plausible that the weaklings are attempting to reduce uncertainty through the integration of information, whereas the superstars may have reduced the need for such systems by reducing the inherent complexity in the factors of production. Of course, an increased emphasis on information systems may in itself create more complexity for the weaklings. Our findings support those of Warner (1987) who "concludes that information technologies should be considered only after conventional improvements and system reorganization have been exhausted."

Using the results of discriminant classification table analysis, we estimated the predictive power of the six discriminating variables. Overall, the independent variables as a group performed 30 percent better than chance; 65.2 percent of the businesses were correctly classified. The function appeared to be slightly better at "hitting" the weaklings, where a 75.3 percent hit rate was observed, or almost 40 percent better than chance . On the other hand, 53.3 percent of the superstars (or about 16 percent better than chance) were "hits."

Manufacturing strategy contents appear to be related to manufacturing strength, but there is clearly much more to winning performance. The data suggest that manufacturing strategy contents (or lack thereof) are more likely to predict weaklings. Predicting the superstars requires that we go beyond the strategy contents of the action programs covered in the survey.

Notably, these findings give credence to the predictive power of the Roth (1989) research linking manufacturing strategy and performance using the 1987 MFP database. In that study Roth stated:

"two manufacturing contents variables are more like to be related to improved manufacturing performance: the degree of emphasis on advanced process technology and attention to manufacturing resources ...Furthermore, it appears that focusing upon advanced process technology, alone, does not appear to be a panacea for manufacturing's malaise; it must be accompanied by a well developed infrastructure. In fact, attention to manufacturing resources is likely to be a manufacturing 'basic' upon which the health of the manufacturing environment rests."

CONCLUSION

Using a manufacturing strategy framework, this paper has empirically addressed the issue of manufacturing and managerial success and business unit performance. Our results, while not conclusive, give tentative empirical support for a growing body of literature on the importance of the manufacturing function as a competitive weapon. Managerial performance was associated with economic outcomes after controlling for the size of the business. Winners, defined on the ability of their managment to meet their goals on qualitative performance measures showed higher ROAs and profitability as a percent of sales than losers. More importantly, managerial winners appear to be associated with heightened manufacturing strength on a number of manufacturing performance criteria, including proxy variables for delivery, flexibility, price, and market scope.

Operational definitions of manufacturing strategy contents were determined, comprising seven important strategy domains. Furthermore, based upon critical manufacturing performance criteria of quality, delivery, flexibility, price and market scope, manufacturing business units were classified into "superstars" and "weaklings," descriptive of their simultaneous relative strength on all manufacturing performance criteria. The manufacturing superstars observed in our study may be representative of the Hayes and Wheelwright (1984) Stage IV performers.

Five of these strategy content areas revealed 30 percent better than chance predictive power in classifying manufacturing superstars and weaklings. Our exploratory analysis suggests that manufacturing superstars are simultaneously placing more emphasis on resource improvements, quality programs and advanced manufacturing technology than the weaklings. On the other hand, the current manufacturing weaklings are expending more effort on restructuring and integrating information and systems with most likely the hope of radical improvement in the future.

REFERENCES

Abernathy and Hayes, "Managing Our Way to Economic Decline," *Havard Business Review*, July-August 1980, 67-77.

Chase, Richard B. and David A. Garvin, "The Service Factory," *Harvard Business Review*, July-August 1989, 61-69.

Cohen, S.S. and J. Zysman, *Manufacturing Matters*, Basic Books, New York, 1987.

DeMeyer, A., J. G. Miller, J. Nakane and K. Ferdows, "Flexibility: The Next Competitive Battle," *Strategic Management Journal*, 10:2,1989, 135-144.

Ferdows, K., J.G. Miller, J. Nakane, and T. Vollmann, "Evolving Manufacturing Strategies," *International Journal of Operations and Production Management*, January 1987.

Fine, C.H. and A.C. Hax, "Manufacturing Strategy: A Methodology and an Illusion," *Interfaces*, 15:6, 1985, 28-46.

Gerwin, D. and J.C. Tarondeau, "International Comparisons of Manufacturing Flexibility, in *Managing International Manufacturing* , (K. Ferdows, ed.), North Holland, Amsterdam, 1989, 169-186.

Green, P. *Analyzing Multivariate Data*. Dryden Press, Hinsdale, IL, 1978.

Hayes, R.H., S. C. Wheelwright and Kim B. Clark. *Dynamic Manufacturing*. The Free Press, New York, 1988.

Hayes, D.R. and S.C. Wheelwright, *Restoring Our Competitive Edge*, John Wiley, New York, 1984.

Hill, T., *Manufacturing Strategy: Text and Cases*, Irwin, Homewood, IL, 1989.

Huete L. and A. V. Roth, "Linking Manufacturing Capabilities with SBU Strategic Directions," *Proceedings of the 1987 Decision Sciences Institute*, 1987.

Lawrence, P. R. and J. W. Lorsch, *Organization and Environment*, Irwin, Homewood, IL, 1969.

McGee, J. and H. Thomas "Strategic Groups: Theory, Research and Taxonomy," *Strategic Management Journal*, 7, 1986, 141-160.

Milligan, G.W. and M.C. Cooper, "An Examination of Procedures for Determining the Number of Clusters in a Data Set," *Psychometrika*, 2, 1985, 159-179.

Miller, J.G., "Report on the 1982 Manufacturing Futures Survey," Manufacturing Roundtable Research Report Series, Boston University, 1982.

Miller, J.G. and W. Hayslip, "Implementing Manufacturing Strategic Planning." *Planning Review*, July-August 1989.

Miller, J.G., and A.V. Roth, "Manufacturing Strategies," *Operations Management Review*,, 6:1, 1988a, 8-20.

Miller J.G. and A. V. Roth, "Manufacturing Strategies: Executive Summary of the 1988 North American Manufacturing Futures Survey." Manufacturing Roundtable Research Report Series, Boston University, 1988b.

Miller J.G. and A.V. Roth, "A Taxonomy of Manufacturing Strategies." Manufacturing Roundtable Research Report Series, Boston University. Presented at the annual meeting of the Strategic Management Society, 1989.

Miller, J.G., A. Amano, A. DeMeyer, K. Ferdows, J. Nakane, A. V. Roth, "Closing the Competitive Gaps," in *Managing International Manufacturing*, (K. Ferdows, ed.), North Holland, Amsterdam, 1989, 153-168.

Richardson, P.R., A.J. Taylor, and J.R.M. Gorden, "A Strategic Approach to Evaluating Manufacturing Performance," *Interfaces*, 15, Nov-Dec 1985, 15-27.

Roth, A. V., "Linking Manufacturing Strategy and Performance: An Empirical Investigation," Manufacturing Roundtable Research Report Series, Boston University. Presented at the National Meeting of the Academy of Management, Washington D.C., August 1989.

Roth, A. V., A. DeMeyer, and A. Amano, "International Manufacturing Strategies: A Comparative Analysis," in *Managing International Manufacturing*, (K. Ferdows, ed.), North Holland, Amsterdam, 1989, 187-211.

Roth, A.V. and J.G. Miller, "1988 North American Manufacturing Futures Survey Fact Book," Manufacturing Roundtable Research Report Series, Boston University,1988.

Roth, A.V., "Differentiated Manufacturing Strategies for the Competitive Advantage: An Empirical Investigation," Manufacturing Roundtable Research Report Series, Boston University. Presented at the National Meeting of the Decision Science Institute, 1987.

Skinner, W., "Manufacturing: Missing Link in Corporate Strategy," *Harvard Business Review*, May-June 1969, 136-145.

Skinner, W. *Manufacturing in the Corporate Strategy*. John Wiley, New York, 1978.

Skinner, W., *Manufacturing: The Formidable Competitive Weapon*, John Wiley, New York, 1985.

Schmenner, R.W. "Behind Labor Productivity Gaps in the Factory," *Journal of Manufacturing and Operations Management*, 1, 1988, 323-328.

Schroeder, R.G., J.C. Anderson and G. Cleveland, "The Content of Manufacturing Strategy: An Empirical Study," *J. Operations Management*, 6, August 1986, 405-415.

Swamidass, P. and W. Newell, "Manufacturing Strategy, Environmental Uncertainty, and Performance: A Path Analytic Model," *Management Science*, 33, April 1987, 509-524.

Ward, P., J. Miller and T. Vollmann, "Mapping Manufacturers Concerns and Actions Plans," *I nternational Journal Of Operations and Production Management*, Fall 1988, 5-18.

Warner, T.N., "Information Technology as a Competitive Burden," *Sloan Management Review*, Fall 1987, 55-61.

Wheelwright, S.C. and R.H. Hayes, "Competing Through Manufacturing," *Harvard Business Review*, January-February 1985, 213-223.

Wheelwright, S. C., "Reflecting Corporate Strategy in Manufacturing Decisions," *Business Horizons*, February 1978, 57-66.

Wheelwright, S.C., "Manufacturing Strategy: Defining the Missing Link," *Harvard Business Review*, January-February 1985, 57-66.

Wheelwright, S. C., "Strategy, Management and Strategic Planning Approaches," *Interfaces*, 14:1, 1984, 19-33.

12

MARKETS, MANUFACTURING STRATEGY, AND TECHNOLOGY ACQUISITION: AN INTEGRATIVE, DEVELOPMENTAL APPROACH

Michael C. Burstein

School of Business Administration
The University of Michigan
Ann Arbor, MI 48109-1234
(313) 936-3248

Pearson Graham
Director (retired)
Corporate Capital Investment, TRW
46025 Mather Lane
Chagrin Falls, Ohio 44022
(216) 247-8563

ABSTRACT

Technology acquisition in manufacturing by U.S. firms during the 1970s and 1980s was characterized often by a frantic grabbing at quick fixes to regain international competitiveness. Unfortunately, such efforts tended to lack any systematic connection with the requirements of specific product-market niches. This paper brings together the contributions of several disciplines to present a customer-driven approach for the acquisition of new manufacturing technology that has manufacturing strategy as its linch-pin. Beyond describing the approach, this paper provides an illustration generalized from several actual cases. The approach is designed to provide a logical flow of information through a series of related tasks that could be expressed by the possibly unique organizational processes of any particular manufacturing firm.

INTRODUCTION

In the frantic activity among U.S. manufacturers recently to recover their competitiveness, desperate leaps at a confusing welter of three letter solutions such as CIE, CIM, JIT, CAM, and TQC have tended to characterize the behavior of many if not most. The chaos wrought by these attempts at the business unit level to adopt the latest vogue in manufacturing management has been amplified further by the patchwork quilt of hardware/software decisions within individual plants often due to a desire like that of business level counter-parts to experience the benefits of new technology.

Unfortunately, such incoherent efforts at the upper reaches of the company hierarchy have tended to lack any specific relationship to customer requirements and to the relative success of the firm vis-a-vis competitors in meeting them. Similarly, the patchwork quilt of technology acquisitions within plants has tended to contribute little or nothing so far as improvements in the manufacturing <u>system</u> performance of these plants. Even where improved performance has occurred, this often has happened in areas with little or no significance for the contribution of the plant to the competitiveness of the firm.

Oddly enough, tools for the linking of a manufacturer's behavior to the exacting requirements of customers from various product-market niches have been available for some time. However, these tools have not been applied together as an integrated whole previously. This is probably due to the variety of disciplines from which the tools come. Drawing upon the discipline of strategic planning, we have systematic procedures (for example, see Hofer and Schendel, 1978) for the determination of appropriate competitive advantages to support the firm in specified product-market niches.

Another discipline, that of technology investment analysis, gives us the means (one source here is Baron, Burstein, and Graham, 1988) to translate choices of distinctive competences into developmental programs for the firm. Meanwhile, the discipline of socio-technical analysis allows us to identify organizational processes within any particular firm (see Ettlie, 1988) for the execution and outcome implementation of formal planning tasks. Perhaps central to the resolution of manufacturing-based problems, the discipline of manufacturing strategy has devised conceptual frameworks (see Van Dierdonck and Miller, 1980) for various distinctive competences at the plant and higher levels in pursuit of competitiveness.

The intent of the authors in this paper is to bring the contributions of these disciplines together in the form of a systematic approach for the determination of appropriate distinctive competences in the manufacturing area of individual firms and, further, for the guidance of technology adoption to realize these distinctive competences. A particular debt of gratitude is owed by the authors to the work of Wickham Skinner (1985) where the connection between factory focus in terms of market requirements on the one hand and distinctive manufacturing competences on the other was explicitly recognized. Beyond describing the general architecture of such an approach, we will provide an illustration of its application to give practitioners and academicians alike a better idea of operational implications.

For purposes of this paper, a manufacturing strategy will be defined as the distinctive competences in manufacturing and in the relationships of manufacturing to the other activities of the firm that have been chosen by the firm to support its competitiveness. Thus, the systematic approach below is designed to facilitate the formulation of manufacturing strategy in the sense of this definition. Another point of clarification, we wish to distinguish at the outset between the concept of distinctive competence and the concept of technology. A capability (e.g., simultaneous engineering) is the nature of a distinctive competence. Several technologies or ways of doing things (e.g, group technology, CAD, CAM, and DFA) might be used in concert to realize a single distinctive competence. Conversely, a single technology (e.g., CAM) might contribute to the realization of several distinctive competences (e.g., simultaneous engineering and product flexibility at the plant level.

THE PROPOSED APPROACH

We will begin the discussion of this section by listing the task sequence that comprises the proposed approach. Then, the embedded architecture of the approach will be described. At this point, we will turn to a discussion of each task in order of sequential appearance with attention to its function, position in the sequence, relation to the overall architecture, and disciplinary origins. The task sequence follows:

1. internal preparation by the business unit
2. determination of product-market niches
3. environmental scan over niches
4. identification of niche-based performance measures
5. estimation of competitor performance by niche
6. assessment of manufacturing responsibility
7. determination of distinctive competences for the plant network
8. assessment of change readiness
9. formulation of network developmental options
10. preparation of a qualitative developmental tree
11. quantitative analysis with tree guidance
12. internal plant preparation
13. determination of plant niches
14. environmental scan over niches of the plant
15. identification of niche-based performance measures for the plant
16. estimation of competitor performance by niche
17. determination of distinctive competences for the plant
18. assessment of change readiness for the plant
19. formulation of plant development options

20. preparation of a qualitative developmental tree for the plant
21. quantitative analysis with tree guidance

Even a casual reading of the task sequence reveals the parallels between tasks 1. - 11. on the one hand and tasks 12. - 21. on the other. Essentially, the former subsequence is macro in focus while the latter subsequence is micro (i.e., plant level) in focus. This macro-micro split is a basic characteristic of the underlying architecture. Note that the early subsequence 1. - 6. really belongs to business-level strategy and sets the stage for the manufacturing staff from the business unit to take the initiative, in collaboration with other functional areas, for identifying an appropriate developmental path at the level of the plant network. It is likely that only early steps in such a path would be implemented prior to a later updating of tasks 1. - 11. with the previous results as starting point. Tasks 12. - 21. could be done by plant employees or by manufacturing staff at the business unit level.

In fact, policy as to the locus of execution for these tasks is an important infrastructural issue for the business unit. This brings us to the next characteristic of the underlying architecture, structure versus infrastructure. In the sense of Hayes and Wheelwright (1984), "structure" refers to the actual processing of materials into product while "infrastructure" refers to all of the activities within manufacturing in support of materials processing. Broadly construed, "infrastructure" even includes interfaces of manufacturing with other functional areas (e.g., product design, marketing, distribution, etc.).

A macro-level example of structure would be supplier-customer relationships between the captive plants of a manufacturing firm. Usually, vertical integration/make vs. buy are language of the supplier-customer aspect of structure at the network level. Hardware/software standards across the plant network often are needed both for structure and for infrastructure. Although neither technologies nor distinctive competences fall neatly into structure/infrastructure as a partition, this categorization is helpful in many tasks of the overall sequence for the initial, roughcut generation of possibilities and for thinking-out the implications of these possibilities in the realm of activities from each of structure and infrastructure.

For purposes of brevity and because of their parallelism, similar tasks from the subsequences 1. - 11. and 12. - 21. will be discussed together. Thus, tasks 1. and 12. both address the synthesis of organizational processes by which the succeeding tasks will be carried out. This synthesis usually requires an analysis first of organizational values and decision-making modes, formal and informal. A major distinction between tasks 1. and 12. is the relative freedom of the business level unit versus the constraints on the plant from business unit policy. Discipline-wise, we owe the background of research for these tasks to socio-technical design (see, for example, Roitman, Liker, and Roskies, 1987). Tasks 2. and 13. require the identification not only of current product (i.e., technology) and market (i.e., customer type) combinations of the firm but also potentially feasible crosswalk opportunities in terms of future product-market combinations not currently within the portfolio of this firm or perhaps any firm. Again, the plant network likely is less constrained than the individual plant. However, the movement today (Lorange, 1980, was one of the early observers of this) away from captive plants as cost centers towards an entrepreneurial stance for such plants can provide some intriguing opportunities outside their traditional role as internal suppliers of parts or subassemblies. The emergence of Fleetguard and Atlas Crankshaft divisions at Cummins Engine is a case-in-point. The identification of product-market niches enables the execution of succeeding tasks differentially by niche. For example, performance requirements for manufacturing often differ from niche to niche; and this has implications for determination of distinctive competences and technologies. However, the flip side of such differentiation is a need to watch carefully for niches that have minimal operating distinctions among them. These might well be grouped together in new "superniches." The discipline of marketing research (Urban and Hauser, 1980, is a particularly good source here) is probably the closest home in the literature for tasks 2. and 13.

The next pair of tasks looks at the suppliers, the competitors, the providers of complementary goods, the relevant governmental players (i.e., local, state, and federal agencies), etc., current or potential, that can affect the outcomes of actions on the part of the business unit in the case of task 3. or of the plant in the case of task 14. Since the focus of the business is principally finished goods, its results from task 3. are of interest to any captive plant in so far as the latter is a potential contributor to the finished goods. However, external markets for the plant pose another distinct focus from that of the business unit since the former frequently has greater sensitivity to intermediate goods opportunities. Environmental scans set the stage for later explanations of competitive position and options for improving position. The strategic planning literature (Michael Porter's work, 1980, is useful here) provides a discipline-orientation for tasks 3. and 14.

The next pair of tasks, 4. and 15., are devoted to the determination of the performance criteria by which the customers in each niche judge the performance of the firm. These criteria have come to be known as critical success factors (CSFs), but the notion of a CSF has found its way into so many contexts at wide variance to this application as to render it useless without additional definition. Once such criteria have been identified, then quantitative measures of them either in manufacturing terms or capable of mapping into manufacturing terms are necessary (Hauser and Clausing provide a particularly explicit illustration of this in their 1988 paper).

Beyond their significance for the planning process, it is the experience of the authors that these performance measures in their effective implementation are the key to the execution of the other recommendation from the planning process. The performance measures provide a basis for assessing the positioning of the firm in relation to competitors. Explanation of relative positions for the manufacturers in a niche then provides a starting point in the determination of distinctive competences. Tasks 4. and 15. are different in the finished goods character of business unit niches versus the intermediate goods nature of niches for individual plants other than final assembly. The disciplines in support of these tasks include industrial engineering (see, for example, Canada, 1986) and market research, the latter for customer criteria, the former for the mapping of manufacturing measures.

Working from previously-defined performance measures, tasks 5. and 16. draw upon survey and/or panel data from customers to estimate the perceived performance of each manufacturer in current niches. Potential niches through crosswalks can be studied similarly if these niches already exist for other firms. Actually, simple examination of current customer perceptions is insufficient for planning, even in the case of current niches. Extrapolation on the basis of changing technology and evolving customer preferences is required to determine the relative position of the firm by the time that already established commitments of the manufacturer to change are played out. The need for such extrapolation in the case of currently non-existent niches from panel data is especially important. Like that between tasks 4. and 15., the contrast for tasks 5. and 16. lies in the distinction between finished goods niches and intermediate goods niches. Again, the relevant disciplines are industrial engineering for measurement of manufacturing system behavior and market research for measurement of customer preferences.

Task 6. has no counterpart at the micro level. This task requires the business unit to allocate responsibility for improved performance among the various functional areas. In the past, manufacturing staff for the business unit were likely to emerge from this process with a mandate strictly in terms of cost reduction with little regard for other product-related characteristics of interest to the customer. Enlightened firms (e.g., D.E.C., Boeing, John Deere) today assign responsibility to manufacturing for improved competitiveness over a broad range of performance measures. Perhaps the discipline most closely associated with this allocation of responsibility is strategic planning.

The remaining tasks (7. - 11.) at the business unit level primarily are the responsibility for the manufacturing function. In the sense of our earlier definition, tasks 7. and 17. result in the formulation of manufacturing strategy. Beginning with the explanation of performance in relation to the competition by niche, the identification of existing and needed distinctive competences is crucial in the overall approach so far as the exchange of information between macro and micro-level planning. Particularly, the business unit people need to know whether to seek competences through network characteristics or through the individual plants. For example, product mix flexibility might be achieved through a range of small plants each assigned responsibility for one product family or through a single large plant with extensive programmable automation.

Similarly, the planning of micro-level competences needs to be informed by policy for the plant network. Determination of distinctive competences can find assistance from the discipline of manufacturing strategy.

Like tasks 1. and 12. before them, tasks 8. and 18. are mostly infrastructural in character. Is each individual employee prepared to take full advantage of the experience and stamina of the organization in making the usually wrenching transition to a new mode of manufacturing operation? Such preparation by the firm often requires extensive training and reorganization (Roitman, Liker, and Roskies provide a good case-in-point in their 1987 paper) that frequently dwarfs the costs of corresponding investments in facilities and equipment. However, the experience of successful manufacturers in recent years suggests the value of such preparation. Once organization processes for change were underway at these firms, they were able not only to accomplish original objectives but had a continuing organizational imperative for further performance improvement. The manufacturing evolution at Hewlett-Packard (Walleigh and Sepehri, 1986) provides an illustration of this phenomenon. The location of these tasks in the approach facilitates the inclusion of such considerations in the formulation of options and developmental paths by later tasks.

Although similar expertise is applied at the macro and micro levels for the execution of tasks 8. and 18., the business unit coordination across functional areas in the plant network can be anticipated to be more complex than analogous activities at the micro level under a decentralized process of planning for the individual plants. The volume of change at the plant level for an extensive factory network can more than even out the resource requirements for change though between the macro and micro levels. The discipline of sociotechnical analysis can play a vital role in the elaboration of infrastructural dimensions for the realization of distinctive competences (see Ettlie and O'Keefe, 1982).

Tasks 9. and 19. call for the translation of desired distinctive competences at the macro and micro levels respectively into technological options capable of performance modelling (e.g., computer simulation) and financial assessment. Here, "technology" is used in the classical sense of industrial engineering to comprehend the way in which men, machines, and materials interact to accomplish something. At the level of the plant network, an infrastructural option might be the transfer of component design engineers from a central office at the business unit to the plant locations of manufacture for the various components. A structural option for a manufacturer of materials (e.g, fiber glass sheets) might be the acquisition of plants for the custom forming of materials into special shapes via infrared ovens. Guidance in the translation of distinctive competences is available through the linking of several disciplines including manufacturing engineering, manufacturing strategy, and socio-technical analysis.

Once developmental options have been generated, these should be joined with possible outcomes under various market conditions, technological events, and governmental activities outside the firm to form alternative multi-year sequences of manufacturing experience for the business unit or plant in the case of task 10. or 20. respectively. Without a sense of sequence, later quantitative analyses will lack specifications, for example, of product mix proportions and overall volume at any time within the planning horizon. Likewise, the incremental effects of successive technological introductions will be virtually impossible to capture. The authors have found decision trees to be the most natural vehicle for the assembly of such developmental paths. Technology investment analysis is a discipline that is focused on the assembly of technological paths (for example, see Graham and Huber, 1988) for manufacturing and product development. Since plant-level trees are likely to incorporate business unit mandates for the plant network, task 10. data might well be helpful to the completion of task 20.

The final tasks of the macro and micro subsequences use the qualitative trees as roadmaps in the modelling of physical performance for various network configurations or plant options under externally-imposed conditions of operation. Then, resulting estimates of physical performance can be translated with the help of additional economic data such as unit costs for infrastructural activities and output prices into time-phased cash flows. These dollar figures can be appended to the qualitative tree in order to do various discounted cash flow calculations and risk analyses. Again, technology investment analysis is a relevant discipline for methodological reference. A point to emphasize on the implementation of the overall approach! The organizational processes by which the tasks above will be executed by any particular firm can be anticipated to depend on the values and organizational forms of that firm. The object of the game is to gain "ownership" throughout the company of results so that the implementation of these results will be facilitated and so that the results represent the best thinking available from the company.

AN ILLUSTRATION OF APPROACH APPLICATION

This fictitious illustration is based upon the experience of the authors with supplier firms to major O.E.M.s from the U.S. transportation industry. For the sake of simplicity, we will assume Midwest Diesel Engine Company (MDEC), the subject of our illustration, to produce a single series of medium size diesel engines for purchase by manufacturers of smaller tractor-trailer trucks. Figure 1. displays the plant network at MDEC. From the information flows for control and feedback in the figure, all staff functions are concentrated at the corporate level. Also, each plant is measured as a cost center.

Figure 1. Midwest Diesel Engine: Macro View

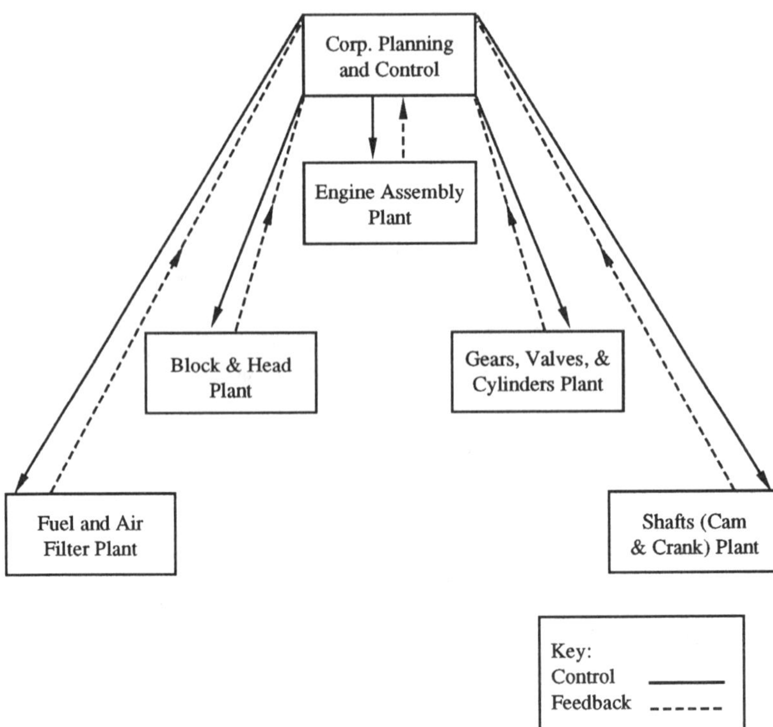

Challenged by a falling volume of orders for its engines during the peak of the most recent business cycle, MDEC has responded with a strategic planning exercise along the lines of tasks 1. - 21. above. Because of the threat to the entire company, each small plant staff along with its local union leadership has been enlisted as a working group in this planning effort. However, the traditions of MDEC place most responsibility for completion of all planning tasks firmly in the hands of the corporate people with plant groups primarily in a data gathering capacity. Moving towards a revival of its order volumes, the corporate staff has identified a possible crosswalk to a niche for larger diesel engines. Some regulatory concerns have been raised by the environmental scan for this crosswalk niche. Apparently, tougher emission standards are anticipated for all truck diesels; and the larger engines present a particularly difficult problem for emissions control. However, the new niche really represents a natural evolution for the current engine series at MDEC for which the company has some substantial technological advantages over potential competitors. Shorter product introduction times and the faster delivery of orders for the old and new lines simultaneously appear to be uppermost in the minds of customers. Quantitatively, this calls for a 40% reduction in times to adjust product mix proportions, a 35% reduction in the cycle time for engineering changes, and a 30% reduction in the lead time for new product introduction.

To achieve these performance improvements, the corporate manufacturing staff has decided to emphasize distinctive competences of responsiveness and flexibility within individual plants. Thus, the company is adopting distributed multi-site management so that each plant will be supported in its transition from cost center to investment center. At the same time, most of the corporate staff will be reassigned among the plants. This will give individual plants an enhanced engineering capability for their components or subassemblies and their production processes. Similarly, additional operations and marketing expertise at the plants will help them to deal with their new status as investment centers. Representing the plants in the strategic planning process, the current working groups will be augmented by these former corporate staffers to consider such competences for their locations as simultaneous engineering and flexible manufacturing.

The working groups are aware of CAD/CAM, GT, and programmably-automated manufacture as technological options for realizing the targeted competences. A corporate expectation is that the macro-level reorganization will contribute about a third of the eventual performance improvement while plant level actions will account for the balance. These expectations are based upon rough-cut computer modelling of the plant network under critical assumptions about the implementation of the new managerial approach and some of the technological options. Investment analysis of these expectations has yielded hope for their economic feasibility.

CONCLUDING COMMENTS

In this paper, we have described an approach for the development of manufacturing strategy and its technological expression. The multi-disciplinary and inter-disciplinary character of this approach received special emphasis. We believe that this approach can help manufacturers to focus their acquisitions of technology upon the realization of market-based distinctive competences. As a subject for future research, we would like to see a test of this approach in multi-site manufacturing firms. Beyond the basic question of whether the approach typically leads to improved performance, the significance of various industrial and organizational attributes would be of particular interest.

REFERENCES

1. Baron, J., M. C. Burstein, and P. Graham, "The Planning and Analysis of Manufacturing Investments," in The Tool and Manufacturing Engineers Handbook (fourth edition) as Chapter 3 of Volume V, Society of Manufacturing Engineers, Dearborn, MI., 1988.

2. Canada, J. R., "Non-traditional Method for Evaluating CIM Opportunities Assigns Weights to Intangibles," Industrial Engineering, Vol. 18, No. 3, March, 1986, pp. 66-73.

3. Ettlie, J. E., Taking Charge of Manufacturing: How Companies are Combining Technological and Organizational Innovations to Compete Successfully, Jossey-Bass, San Francisco, CA., 1988.

4. Ettlie, J. E., and R. D. O'Keefe, "Innovative Attitudes, Values, and Intentions in Organizations," Journal of Management Studies, Vol. 19, No. 2, 1982, pp. 163-182.

5. Graham, P. and R. F. Huber, "Guidelines for Better Investment Decisions," Production, Part I (March) and Part II (April), 1988.

6. Hauser, J. R. and D. Clausing, "The House of Quality," Harvard Business Review, May-June, 1988, pp. 63-73.

7. Hayes, R. H. and S. C. Wheelwright, Restoring Our Competitive Edge, John Wiley & Sons, New York, NY., 1984.

8. Hofer, C. W., and D. Schendel, Strategy Formulation: Analytical Concepts, West Publishing Co., St. Paul, MN., 1978.

9. Lorange, P., Corporate Planning: An Executive Viewpoint, Prentice-Hall, Inc., Englewood Cliffs, NJ., 1980, pp. 80, 81.

10. Porter, M. E., Competitive Strategy: Techniques for Analyzing Industries and Competitors, The Free Press, New York, NY., 1980.

11. Roitman, D. B., J. K. Liker, and E. Roskies, "Birthing a Factory of the Future: When is 'All at Once' Too Much?" in Corporate Transformation, R. H. Kilman, T. J. Covin and Associates, Jossey-Bass, San Francisco, CA., 1987.

12. Skinner, W., Manufacturing: The Formidable Competitive Weapon. John Wiley & Sons, New York, NY., 1985.

13. Urban, G. L. and J. R. Hauser, Design and Marketing of New Products, Prentice-Hall, Inc., Englewood Cliffs, NJ., 1980.

14. Van Dierdonck, R. and J. G. Miller, "Designing Production Planning and Control Systems," Journal of Operations Management, Fall, 1980.

15. Walleigh, R. and M. Sepehri, "H-P Division Programs Reduce Cycle Times, Set Stage for Ongoing Process Improvements," Industrial Engineering, Vol. 18, No. 3, March, 1986, pp. 74-81.

13

PILOT PLANTS TO ASSIST INDUSTRIAL FMS APPLICATIONS

Géza Haidegger - Dr. George L. Kovács

Computer and Automation Institute, Budapest, HUNGARY
H-1502, POB. 63. Phone: 36-1-1811-143; Fax: 36-1-1667-503

INTRODUCTION

The past years of industrial development proved that computer aided techniques and technologies, as Computer Aided Design (CAD), Computer Aided Manufacturing (CAM), Computer Aided Process Planning (CAPP), etc. will be used to get sophisticated, manufacturing systems of high level of automation. The so-called Integrated and /or Flexible Manufacturing Cells (IMC,FMC) and Flexible Manufacturing Systems (IMS, FMS) are getting to be more and more spread out world-wide and they intend to be real Computer Integrated Manufacturing (CIM) systems.

The cooperation of different computers with different numerically controlled (NC) machines and with each other is achieved by means of computer networks with various protocols. The Manufacturing Automation Protocol (MAP) offers possibilities to give standardized solutions to most crucial interconnection questions.

In the first part of our paper a short account will be given on Hungarian R&D and implementation activities in the field of industrial networks [1,2,4]. The second part of our paper will discuss some efforts to get industrial people involved and to promote education of factory automation, including MAP/TOP. These activities will be supported by means of two sophisticated CIM pilot plants, one at the Technical University of Budapest (TUB), the other one in the Computer and Automation Institute (CAI).

The pilot plants will be used not only by academicians, but by industrial people as well to test their design and production needs and possibilities and to make experiments on different manufacturing strategies. This way engineers and managers can get information on future manufacturing systems not only by designing them, but by making some tests on similar purpose systems.

INDUSTRIAL NETWORKS AND CIM PILOT PLANTS IN HUNGARY

Not only research institutes and technical universities, but industrial firms and factories started to develop networked computer controlled systems almost ten years ago.

- Five different FMS were developed in factories - with the assistance of university and research institute R&D experts - with different levels of automation.

- In the field of process control a PROWAY-type network was developed to be used in gas, oil, steel, sugar and cement industrial plants.

- PDP-type minicomputers were connected into nets to control certain elements of nuclear power plants and to realize measurements and their evaluations.

- Control of discrete parts manufacturing was supported by means of networked high-end PLCs. Plastic industry used the same type of networked PLCs to control extrusion processes.

- A multipurpose, distributed PLC module-system was developed here, and then after license agreements sold to different western countries for manufacturing and marketing.

- A special, coax based LAN was designed and implemented to connect all different computers of our institute and to give us a gateway possibility to reach other, remote computer systems. This same kind of office network is used in 8-9 other companies in Hungary, GDR and USSR.

Some weak points of the above R&D efforts and results became clear fast enough :

* As the developments were neither coordinated nor any standards were taken into account, the results became not compatible when such demands arose.
* The systems were not upgradable, thus further developments were difficult or even impossible in several cases.

New, centrally supported projects were started three years ago to develop new families of controllers of different levels with the direct goal of being open and compatible with the international standards. Such a module system was to be designed which matches the aims of an overall CIM concept [1].

The R&D efforts were supported by purchasing internationally accepted licenses as well to speed up the developments and implementations and to guarantee international compatibility.

Our institute (CAI) is deeply involved in the above R&D projects, and among others we developed a 16-bit industrial cell-controller. The development of the cell-controller was coordinated with all other Hungarian controller developments (PLC, CNC, RoC, etc.). A standard IBM PC was programmed in C to solve all tasks of a manufacturing cell controller. The PC is enhanced by an intelligent coprocessor module card to ensure fast communication with all elements (CNCs) of the cell. Even a production scheduling system belongs to the tasks of the cell-controller [5,6].

Two, government supported CIM pilot plants are under development, which plan to use the most up-to-date MAP, Ethernet and other networking technologies [7].

One of the main reasons of supporting these pilot plants is to promote industrial FMS/CIM investments by decreasing the recently wide gap between research/development experts and factory engineers. It is a real problem for R&D people to convince the possible users (factory personnel) to apply the available R&D results in industry.

Both plants should be able to demonstrate all available (Hungarian and foreign) CAD, CAM, etc. programs and technologies to assist industrial people to get the necessary knowledge based on which they will be able to cooperate with R&D (and with foreign trade) people in the design and implementation of up-to-date computerized plants (workshops).

The possibility of checking (simulating/modeling) some main features of future investments seems to be attractive not only for technical people, but for managers as well.

The pilot plant of the Computer and Automation Institute will be a CAD-oriented CIM plant containing 32-bit computers and graphics workstations and a manufacturing cell. The CAD stations and the 16 bit cell-controller are connected by Ethernet network (see Fig. 1.). The manufacturing cell consists of a 5-axes CNC milling-machine, a CNC lathe, a CNC measurement machine and a robot and further elements will be added in the future [11].

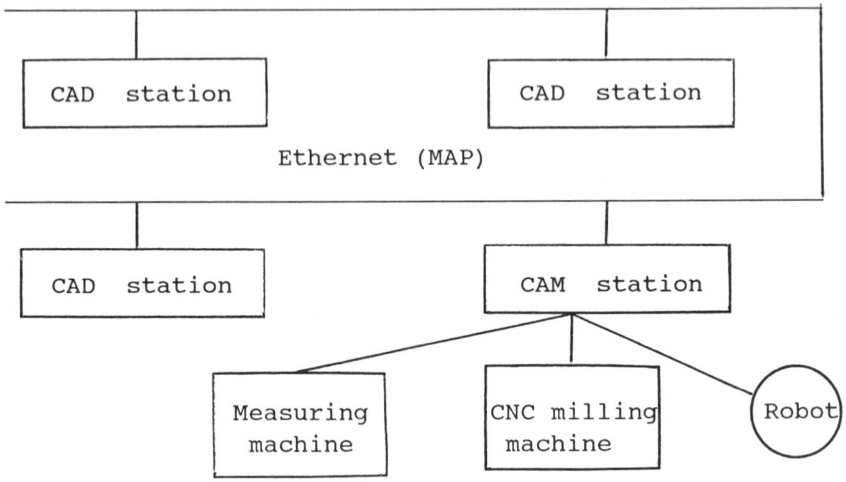

Fig. 1. CAD-oriented CIM system

Fig. 2. shows the CAM-oriented part of the computer network of the Technical University of Budapest. The whole system consists of 32-bit and 16-bit computers in seven different buildings, all are connected by Ethernet. In each building there are one or more LANs connecting the PCs within the buildings. The 32-bit computers run the CAD-programs, and all other machines can be connected via the network to be terminals.

The CAM-oriented CIM subsystem is located in one of the workshops, and it has its own 32-bit host, which is connected to all 16-bit cell-controllers. The LAN of the CAM system is planned to be a MAP 3.0 network, as soon as possible [12].

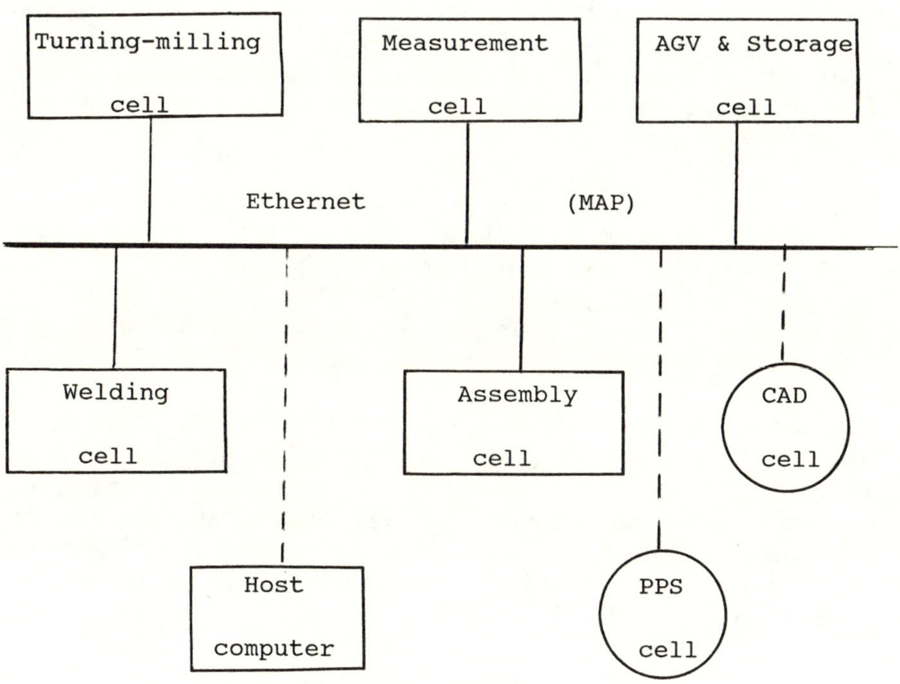

Fig. 2. CAM-subsystem at the TU of Budapest

THE MAIN OBJECTIVES OF THE PILOT PLANTS

The basic goals of both systems are the same :

- To improve the research-development-education possibilities, to support the efficacy of work in both institutions (TUB and CAI).

- To support unification (standardization) on the Hungarian CAD/CAM scene. This should be done in accordance with the international standards from the point of view of systems and system elements as well, using as much 'home-made' (Hungarian) elements as possible.

- Access should be provided to all Hungarian and available foreign CAD/CAM programs. This means access to all precise, up-to-date documentation too. Interfaces should be provided between the programs within the systems and to programs which are not yet integrated.

- The systems should always be ready to demonstrate all the new (and older) R&D results for internal and external (academic and industrial) experts.

- The systems should be used for graduate and post-graduate education, too.

The systems should provide the possibility to investigate the production processes, including turning, milling, grinding, welding, assembly, etc. and to investigate machine-tools, CNC and cell controllers. The running and testing of the available CAD/CAM programs should be provided, too. This way simulation of future CIM investments becomes realistic.

R&D ACTIVITIES ON THE CIM PILOT SYSTEMS

The results and experiences in the Computer and Automation Institute and at the Technical University of Budapest of the past ten years in the CAD/CAM field are the basis of the work which will be done in the following years.

- The development of mechanical engineering design and manufacturing systems, based on surface and volumetric modeling. Research and problem-solving of product modeling.

- Research on different levels of MAP and other networks and in the field of network compatibility. Cell- and system-level implementation of MAP protocols.

- The research and development of control systems of flexible, integrated manufacturing cells and systems, and of robot and machine-tool controllers; system integration studies.

- R&D of automatic measurement and supervision of machine-tools, and machined parts during manufacturing, using computer aided diagnostics; reliability tests and state supervision. Testing of workpieces and tools during cutting processes; coordinate-measurement technique.

- Research of control systems of manufacturing processes, and of integration of different production systems.

- Coupling of manufacturing system control with production planning and other preparation packages.

- Research and development of different fields of attendance-poor manufacturing, as :

  * material handling, technological design , fixtures, palettes, etc.

  * production planning systems, data- and material-flow, etc.

- Application of the methods and means of artificial intelligence, as e.g. knowledge-based and expert systems to the above listed R&D fields.

EXPERIMENTS ON MANUFACTURING STRATEGIES

The hardware-software possibilities of the two pilot plants (including the CAD/CAM/.. programs, networks and mechanical engineering equipment) make it possible to use them according to prescribed design/production conditions.

The given conditions can meet the requirements of planned, future FMS (CIM) implementations, giving the possibility of test-runs and simulation/modeling.

Hierarchical and Structural problems

If we suppose a four level hierarchical FMS organization (CNC maçhine tool, cell, plant/workshop, factory) one has to decide some questions, which may be verified by experiments :

- On which level of the hierarchy should the production planning/process planning (scheduling) be done ?

- On which level (and how) re-scheduling should be processed in the case of any kind of erroneous operation of any element of the system on any of the hierarchical levels ?

- Should the production planning be on-line or off-line?

- On which level (and how) intervention should be done if necessary ?

The correct answers to these and other similar kind of (feed-back) questions assist mostly in proper design of organization of CAD/CAM programming and in evaluating the structure of the planned systems.

Production/manufacturing problems

The greatest part of our future experiments deals with finding optimal (or acceptable) solutions of manufacturing according to given tasks.

There are several studies in literature (e.g. [8], [9], [10]) which give (experimental) data on batch/lot size, flexibility, productivity, number of different products, production volume, necessary man-power, etc. In most cases increasing flexibility decreases the production volume and vice-versa. The decision of using traditional job shop, special automated systems or flexible manufacturing cells/systems, or some combination of them is really crucial.

Generally an FMS producing 100-1000 different products is regarded as an advanced one. This number and all the above data are different according to Western European, USA or Japanese experiences. As we do not have reliable data from the Eastern European countries we plan to use real production tasks to get them. These tests will be the tests of the available software means as well.

As process/production planning is of great importance schedulers of different levels will be tested according to - partly contradictory - criteria, as e.g :

- To minimize the number of products waiting for operation
- To maximize the throughput of the system
- To minimize the average time that products spend in the system
- To achieve JIT operations
- To minimize idle times
- To generate average load on machinig resources

Generally no real optimum can be achieved, but one has to manage with good, acceptable solutions as computing capacity and hardware resources are both restricted.

Flexibility studies

The flexibility and the effects of flexibility on other properties of a system/ system components can be examined by means of experiments, too.

The way of getting flexibility, and the methods of measuring the level of flexibility is known from earlier studies (e.g. [8]).

We intend to deal with the flexibility of the following, which has a strong correspondence to each other :

	- product	
- machine	- process	
	- operation	
		- production
- routing	- volume	
	- expansion	

Some of the relationships of these flexibilities will give us good ideas on how to implement and how to run a future FMS.

The two pilot plants will serve as a background to investigate the above mentioned manufacturing strategies and concepts. The test-runs and evaluation of the gathered data will keep us busy for some time.

CONCLUSION

The world-wide proliferation of highly automated manufacturing systems can be imagined only if the system modules are standardized (see ISO, OSI, MAP, TOP, VME, EIA, DIN, GOST, etc.).

On the other hand, an up-to-date, possibly computer aided design-implementation methodology, such as SATT-DOC [3] should be applied. The working manufacturing systems have proven that the big systems are getting more and more complex, by adding new elements and by changing old elements and functions according to demands, which would not be possible without having standard subsystems and without a good design with a certain 'look-ahead'.

Active cooperation between different countries in R&D and in implementing systems is impossible without respecting all international standards and prescriptions. The two CIM pilot plants, which are recently implemented in Hungary are serving the above goals, by giving companies a chance to use up-to-date software-hardware means and by assisting education on the graduate, post-graduate and factory level as well. The internationally accepted MAP/TOP standards are planned for application.

One of the basic applications of the pilot systems will be the test runs to get information on the behaviour of such systems when driven by various manufacturing strategies. End-users of new FMS installlations are eagerly waiting for results on how to operate future FMS implementations.

References

[1] Dr. L. Nemes, Dr. Gy. Kovács: Research and development of CIM Sytems - National Project in Hungary. FMS Conference, Magdeburg, 1985. ed. in Moskow, 1986. pp. 55-67. (in Russian)

[2] G. Haidegger et al.: Manufacturing Systems based on Distributed Intelligence Cell Concept. IFIP WG Conference on Integration of CAD/CAM. NH. Gaussig GDR.1983 Nov.pp. 253-263

[3] Dr. Z. Létray, G. Haidegger: SATT and MAP as guidelines for CIM implementations. CAD'88 Conference Moskow, March 22-24.

[4] G. Haidegger et al.: Distributed Control System for Machines in Manufacturing. IFAC Symposium on Microcomputer Application in Process Control, Istambul 1986. pp. A6/65-73.

[5] P. Bertók et al.: The Integration of Intelligent Cell Controllers into Factory Networks. PROLAMAT'88 Conference, Dresden 1988. 14-17 June.

[6] G. Csurgai et al.: Intelligent Cell Controllers form Nodes on Factory Network. 17th International Symposium on Automotive Technology & Automation (CIM & MAP). ISATA 87178. Volume 1. 1987. Oct.25-30. Münich, GFR.

[7] G. Haidegger et al.: New trends in the development of MAP. COMPCONTROL'1987 Moskow Conference

[8] J. Browne et al.: Classification of flexible manufacturing systems. The FMS Magazine, April, 1984, pp. 114-117.

[9] M. Bonfioli et al.: Production scheduling and operational control of flexible manufacturing systems. Robotica (1985), vol. 3. pp. 233-243.

[10] H. Bakker : DFMS : A New Control Structure for FMS. Computers in Industry 10 (1988). pp.1-9.

[11] Gy. Hermann et al.: CAD System Description, 1989, Computer and Automation Institute, Mechanical Engineering Automation Division (in Hungarian)

[12] L. Cser et al.: CAD and CAM System Description, 1987, Technical University of Budapest, Information Laboratory and Mechanical Engeneering Department (in Hungarian)

14

MANUFACTURING STRATEGY AND NEW VENTURE ORIGIN: AN EMPIRICAL ANALYSIS

DERRICK E. DSOUZA, University of North Texas
PATRICIA P. MCDOUGALL, Georgia State University
RICHARD H. DEANE, Georgia State University

ABSTRACT

This research addresses key elements in the manufacturing strategy of 64 new venture firms in the computer and communications industries. More specifically, the research is focused on manufacturing strategy differences between new ventures initiated by existing firms and new ventures initiated by entrepreneurs. Questionnaire and personal interview data were analyzed via a discriminant analysis approach. The research reveals that significant differences in manufacturing posture do exist as a function of new venture origin, i.e. there is a statistically significant difference between corporate sponsored and independent new venture firms in terms of key manufacturing strategy variables.

The study suggests that independent new ventures compete through superior product quality, using in-house manufacturing technology. Products are directed to a few large customers, and new product development is not a primary issue. Corporate sponsored ventures are less concerned with superior product quality, but place a greater emphasis on product manufacturing technology, new product development and sub-contracting.

INTRODUCTION

The need for a well formulated and successfully implemented manufacturing strategy is greater than it has ever been. Intense competition, at both the national and international level, has made companies search for approaches that will provide assistance in the development of a competitive advantage in the market-place. Environmental and firm-based dynamics are therefore at levels that compel a firm to initiate and to implement a well conceived manufacturing strategy. Several authors such as Skinner (1978), Hayes and Wheelwright (1984), and Buffa (1984) have pioneered the call for a greater emphasis on operations strategy as one way of creating, maintaining and improving a firm's competitive edge. Numerous theories/models, drawn from strategic management principles, have been developed to represent the process involved in manufacturing strategy (e.g., see Fine and Hax, 1985).

Manufacturing strategy research has focused on established firms, either ignoring new venture firms or failing to treat these firms separately. New venture firms face unique challenges and opportunities in developing a viable operations strategy. While unhampered by the "sunk cost" of existing facilities and equipment, new venture firms often lack the necessary funding and managerial experience to develop and execute a competitive manufacturing strategy. Despite the growing interest in manufacturing strategy and the continuing interest in new venture enterprises, there is little, if any, reported research that identifies the manufacturing posture of new venture firms entering the marketplace.

As such, this research examines key elements in the manufacturing strategy of 64 new venture firms in the computer and communications industries. Of particular interest are manufacturing strategy differences in relation to new venture origin. For the purpose of this research, new venture origin refers to whether the new venture firm was initiated by an existing firm (henceforth called a "corporate sponsored" firm) or by an independent entrepreneur (henceforth called an "independent" firm).

This study examines the manufacturing strategies of corporate sponsored and independent new ventures and assesses the effect the origin of the firm has on its manufacturing strategy.

Manufacturing Strategy

Manufacturing strategy has received increased attention in recent years as an important element in a firm's attempt to gain a competitive advantage in the marketplace (Skinner, 1969; Hayes & Wheelwright, 1984; Adam & Swamidass, 1989). While a firm's overall business strategy essentially outlines the general steps to be taken to attain a set of long range objectives, manufacturing strategy defines the manner in which manufacturing resources are deployed to reach the firm's objectives (Hayes and Wheelwright, 1984). Manufacturing strategy, a sub-component of the firm's overall strategy, must be consistent with the overall strategy of the firm, and also with the strategies of other functional areas (e.g., marketing, finance, personnel, etc.)

Numerous authors have attempted to outline the specific pattern, or types, of decisions and actions that comprise the manufacturing strategy of a firm (see for example, Skinner, 1978, Hayes & Wheelwright, 1984, Buffa, 1984, Ferdow, 1985.) It is this pattern of manufacturing decisions and actions that are critical to the firm's ability to meet strategic goals through manufacturing. Most prior manufacturing strategy "content" research of this type has been subjective and prescriptive in nature. In addition, these researchers have examined the characteristics of manufacturing strategy in larger, established firms (e.g., Ferdows, 1985). Grouped together, this body of research has identified various decision areas for operation strategy.

Although there has been a proliferation of operations strategy literature in the last ten years, relatively few empirical studies have been undertaken. In their recent review of operations strategy research, Adam and Swamidass (1989) identified the use of empirical research as building blocks for operations strategy theory development as one of the missing themes in operations management. In particular, they noted, "Operations strategy research based on PIMS-type (Schoffler et. al., 1974) databases are needed to move forward and approach maturity." (1989, p.193).

New Ventures

New venture firms play a key economic and social role in society. Developing an improved understanding of how and why new businesses succeed is critically important, not only to entrepreneurs starting firms, but is also of growing concern to many existing corporations as they move into new markets or sponsor start-ups to stimulate innovation or enhance growth, as well as to public policy makers who view new businesses as critical to sustaining economic growth and social mobility.

Corporate sponsored and independent new ventures face different challenges when entering an industry (Hines, 1957; Bevan, 1974; Fast, 1981; Heflebower, 1954). Commenting on his research examining the impact of entry strategies on the performance of a group of corporate sponsored new ventures, Biggadike cautioned, "the results probably do not apply to new ventures that are individually launched (1979, p. 104)."

Past research on new venture start-ups indicate several advantages that are often associated with corporate sponsored entry into an industry (Caves and Porter, 1977 and Hines, 1957): (1) availability of unused capacity, (2) economies of scale, (3) access to capital, (4) skilled management, (5) possibility of vertical integration, (6) access to technology and markets, (7) staying power, and (8) synergy with firm's existing products and technologies. Likewise, several advantages are often associated with entry into the marketplace by an independent firm (Fast, 1981, Weiss, 1981, Stocking and Watkins, 1951): (1) enhanced possibility of clear and definite objectives, (2) more of a ends oriented and performance oriented environment, (3) unencumbered by existing process technology, (4) superior motivation of managers, (5) lack of bureaucratic inertia, and (6) clear decision process.

The industrial organization literature strongly suggests that corporate sponsored firms hold market entry advantages over independent firms (Hines, 1957). Such a finding was justified on the basis that certain structural characteristics such as scale, capital and product differentiation act as entry barriers to an industry. Better access to resources should therefore provide the corporate sponsored firm with market entry advantages. However, empirical research specifically addressing this question suggests a different relationship. Weiss (1981) provided one of the first comparative analysis of corporate sponsored and independent new ventures. He found that independent new ventures gained momentum after the second year of operation, reached profitability in half the time taken by corporate sponsored firms to reach profitability, and that by the eighth year, had outperformed their corporate sponsored firm counterparts (30% average ROI for independent firms vs. 7% average ROI for similar corporate sponsored firms).

Entry advantages of either corporate sponsored firms or independent firms are likely to be linked with the manufacturing strategy of the new venture. For example, the strategy literature offers cogent arguments to justify the proposition that independent firms are more likely to enter the market on a niche basis (Broom and Longenecker, 1979; Bucheler, 1967; Katz,1970). Such a niche strategy is normally associated with small scale production facilities that do not compete on the principal basis of cost leadership. By contrast, corporate sponsored firms would be more capable of market entry on a large scale basis to attack the central core of the market (Hines, 1957). Assuming the viability of such an argument, resulting differences in manufacturing strategy should arise logically as a function of new venture origin.

Direct research on manufacturing strategy and the new venture firm is void. This research is thus "ground-breaking" in its attempt to relate, through an empirical study, elements of a firm's manufacturing strategy with new venture origin.

Research Methodology

Sample and Procedure

This research is part of an ongoing, comprehensive study of new venture firm strategy and performance (McDougall, 1987). The research data base was compiled from mail questionnaires, validated in a pilot study, mailed to the owner, president, CEO, or chairman of the board of new venture firms within nine SIC codes in the computer and communications equipment manufacturing industries. The owner, president, CEO, or chairman of the board, was targeted as the respondent because they typically possess the most comprehensive knowledge of relevant characteristics of the organization, the firm's strategy and its performance (Hambrick, 1981). Consistent with prior research, a firm was considered a "new" venture if it was eight years old or less (Biggadike, 1979, Miller and Camp, 1985).

The sample was checked for representativeness by comparing the research data base with the population of all Dun and Bradstreet new ventures in the industry segments studied, based on number of employees and geographical location. A chi-square analysis confirmed that survey respondents did not differ significantly (p = .05 level) from non respondents based on firm size or geographical location. Additionally, a follow up telephone survey of 23 non respondents was conducted.

There were no significant differences in any of the basic characteristics describing non respondents and the research data base.

Reliability was examined using a sample of survey second respondents. Second responses by a different top executive in the firm were received from 50 firms randomly selected to assist in instrument validation. Average correlation scores of 0.58 between the pairs of respondents on the strategy variables gave the researchers confidence in the instrument's reliability.

Table 1
Manufacturing Strategy Categories and Variables Used

Strategy Categories	Variables
* Capacity and facilities	1. Capacity utilization
* Vertical integration (backward)	2. Level of sub-contracting
* Technological focus	3. Public domain versus patented technology
	4. Top management expertise (Manufacturing versus marketing)
* Process focus	5. Proven versus innovative manufacturing process
* Cost	6. Concern for being the lowest cost producer
* Product/service design	7. New product development emphasis and focus
	8. Narrow versus broad product range
* Quality	9. Emphasis on servicable, versus superior product quality
* Service	10. Level of customer service provided
	11. Average customer order size

The data base contains detailed information relative to the business strategy of each firm, to include key variables relating to the manufacturing elements of a firm's strategy. These eleven manufacturing strategy variables have been listed in the right-hand column of Table 1. Based on a review of the literature, each of the variables have been categorized by various generally recognized decision areas for manufacturing strategy. These strategy categories are shown in the left-hand column of Table 1. An example of how 3 of these variables were operationalized on a seven-point, bi-polar scale is presented in Figure 1.

Figure 1
Example of how Manufacturing Strategy Variables were Operationalized

Emphasis On	Neither Extreme Emphasized	Emphasis on
1. Emphasis on serviceable product quality.	|_|_|_|_|_|_|_|	Emphasis on superior product quality.
2. Sub-contracting or sourcing of production.	|_|_|_|_|_|_|_|	Fully integrated production.
3. Continued new product development.	|_|_|_|_|_|_|_|	Maintaining current products.

Pairwise Matching of Corporate Sponsored and Independent Firms

An initial step in the research was to segment the sample into corporate sponsored and independent firms. The entire data base consists of 247 new venture firms. Of these, there were 32 "corporate sponsored" firms for which complete manufacturing data was available.

A "matched pair firms" approach was used to eliminate confounding factors to the extent possible. That is, each of the 32 corporate sponsored new venture firms was matched to an independent new venture firm on the basis of industry segment, age, and number of employees. Four-digit SIC codes were used to define industry segments. These variables, along with the means and standard deviations for each group, are presented in Table 2.

The mean date the firm began operations was 1981.1 for corporate sponsored firms, versus 1981.5 for independent firms. The 61.4 "number of employees" mean of the corporate sponsored group was higher than the 33.8 mean of the independent group. With 209 independent new ventures in the database available for matching with the 32 corporate sponsored firms this large difference was somewhat surprising, although consistent with observations and prescriptions of many new venture practitioners and researchers. If independent new ventures have entered the market in niches as prescribed by many, it would follow that they would in general be much smaller in size than corporate sponsored new ventures entering the same industry on a larger scale.

Table 2
Pairs-Matching Criteria Used

Criterion	Parented		Independent	
	Mean	Std Dev	Mean	Std Dev
Industry Code	(4-digit SIC codes used)			
Year Business Began	1981.1	2.8	1981.5	2.1
Number of Employees	61.4	121.4	33.8	48.9

SIC codes included in the paired sample: 3555, 3573, 3579, 3613, 3652, 3661, 3664

Analysis and Results

To determine whether the manufacturing posture for new ventures is influenced by the origin of the business, a canonical discriminate analysis was performed. Canonical discriminant analysis essentially develops linear composites of the Y variable (type of firm) and the X variables (the eleven operations strategy variables). The resulting squared canonical correlation is analogous to the R-square in multiple regression, providing the amount of variance explained by the canonical function.

Table 3
Standardized Connical Coefficients

Strategy Variable	Standardized Connonical Coefficient
1. Emphasis on serviceable vs. superior product quality	- 0.7289
2. Level of sub-contracting	- 0.5402
3. New product development emphasis	0.5240
4. Average size of customer order	- 0.4295
5. Proven vs. innovative manufacturing process	0.4093
6. Public domain or patented technology	0.3249
7. Narrow vs. broad product range	0.3249

8. High vs. low capacity and facility utilization	0.2985
9. Concern for being the lowest cost producer	0.1980
10. Level of customer service provided	0.1974
11. Top management background expertise	0.1534

The multivariate test of significance of the discriminating power of the eleven manufacturing strategy variables as a group, provided a Wilks' Lambda of 0.634. This corresponds to a "p" value of 0.0074, well below the conventional level of significance of 0.05. The null hypothesis of no difference between corporate sponsored and independent firms is rejected; therefore the corporate sponsored and independent firms are different based on the eleven manufacturing strategy variables.

An analysis of the absolute value of each of the individual canonical coefficients is helpful in determining which of the eleven operations variables are more important in discriminating between corporate sponsored and independent firms. These individual canonical coefficients are presented in Table 3.

Conclusions and Suggestions for Future Research

This empirical research has been somewhat ground-breaking in attempting to relate elements of a new venture's manufacturing strategy to its origin. The results suggest that the manufacturing strategy or posture of corporate sponsored firms tends to be significantly different from independent firms in the computer and communications equipment manufacturing industries.

Different manufacturing strategy profiles for corporate sponsored and independent new venture firms in the computer and communications manufacturing industry are suggested by the results. Table 4 provides the mean and standard deviation of scores of corporate sponsored and independent firms on each of the eleven variables. The table indicates that independent firms tend to attack the market with superior product quality using proven, in house manufacturing technology. Products are directed to a few large customers, and new product development is not a primary issue with the independent firm. If we are to assume that the independent new ventures are small, lower market share firms, the findings that these new ventures employ superior product quality as their competitive edge is consistent with much of the new venture strategy literature prescribing high quality niches for market entry. The lack of new product development is not surprising given that innovating firms must continue efforts on a very large scale to be successful (Drucker, 1985). It is doubtful whether the independent ventures in this sample are of a large enough size that a product innovation strategy could be successfully implemented.

Table 4
Mean Responses on Key Manufacturing Variables*

		Parented		Independent	
		Mean	Standard Deviation	Mean	Standard Deviation
1.	Public domain or patented technology	2.0	1.8	3.9	2.0
2.	Emphasis on servicable v/s superior product quality	4.9	2.0	5.6	1.9
3.	New product development emphasis	3.0	2.0	2.3	1.6
4.	Average size of customer order	4.5	1.9	4.9	2.0
5.	Level of sub-contracting	3.2	2.0	4.1	2.4
6.	Proven v/s innovative manufacturing process	4.4	1.8	3.8	1.9
7.	Level of customer service provided	5.7	1.6	6.1	1.5
8.	Narrow v/s broad product range	3.2	2.2	2.9	2.0
9.	Concern for being the lowest cost producer	4.3	1.8	4.1	1.8
10.	High v/s low capacity and facilities utilization	4.0	1.8	3.8	1.9
11.	Top management background expertise	3.5	2.2	3.0	2.0

* Variables measured on a 1-7 scale.

Table 4 also indicates that in contrast to independent new ventures, corporate sponsored firms indicate less concern with superior product quality, but place a greater emphasis on innovative manufacturing process technology and new product development. Sub-contracting is an integral element of their operations strategy, suggesting that, sub-contracting can be used effectively to manage the sporadic level of customer's orders that could otherwise inhibit aggressive growth objectives. Thus the new venture can meet peak demands, and at the same time avoid the over-capitalization that is frequently associated with the high failure rate of new ventures.

That the survey covered only two types of industries is a limitation in generalizing the results. However, the dynamic nature of these two industries over the past several years would highlight any differences that may exist between the two groups (corporate sponsored and independent firms). In addition, only a limited number of industries have a large enough sample of new ventures to allow the researcher the opportunity to study these issues.

Another possible limitation of the study is the differences in employee size of the corporate sponsored and the independent ventures. It is possible that the profile differences may actually be attributed to venture size rather than venture origin. Future research should attempt to control for size to a greater degree than was possible in this research.

This research provides support that the manufacturing strategy profile of corporate sponsored new ventures can be distinguished from that of independent new ventures. Additional research is needed to further refine these relationships, as well as to examine other manufacturing strategy variables.

Although this research makes a distinction between the manufacturing strategy of corporate sponsored and independent new ventures, it leaves unanswered the question of whether this distinction is important with respect to new venture success. Future research should incorporate a normative perspective, thereby allowing prescriptive recommendations to be forwarded.

References

Adam, E. E. & Swamidass, P. M. (1989). Assessing operations management from a strategic perspective. Journal of Management, 15(2), 181-203.

Bevan, A. (1974). The U. K. potato crisp industry, 1960-72: A study of new entry competition. Journal of Industrial Economics, 22, June 1974, 281-297.

Biggadike, R.E., The risky business of diversification. Harvard Business Review. 57, May-June 1979, 103-111.

Broom, H. N. & Longnecker (1979). Small Business Management. Cincinnati, OH: Southwestern.

Buchele, R. B. (1969). Business policy in growing firms. San Fransisco: Chandler Publishing.

Buffa, E.S. (1984). Meeting the Competitive Challenge - Manufacturing Strategies for U.S. Companies. Homewood, IL: Richard D. Irwin.

Caves, R. E. & Porter, M. E. (1977). From entry barriers to mobility barriers: conjectural decisions and continued deterence to new competition. Quarterly Journal of Economics, 91, May 1977, 241-261.

Drucker, P. F. (1985). Entrepreneurial strategies. California Management Review, 27, Winter 1985, 9-25.

Fast, N. D. (1981). Pitfalls of corporate venturing. Research Management, March 1981, 21-24.

Ferdows K. (1985), "Implementation of Operations Strategy" in Proceedings of the 4th Annual Operations Management Association Meeting. Tempe, AZ, 1985, 57-59.

Fine, C.H. & Hax, A.C. (1985). Manufacturing strategy: A methodology and an illustration", Interfaces, 15(6), 28-46.

Hambrick, D. C. (1981). Strategic awareness within top management teams. Strategic Management Journal, 2, 263-279.

Hayes, R.H. & Wheelwright, S.C. (1984). Restoring our competitive edge: Competiting through manufacturing. New York: John Wiley & Sons.

Heflebower, R. B. (1951). Economics of size. Journal of Business of the University of Chicago, 24, 258-273.

Hines, H. H. (1957). Effectiveness of "entry" by already established firms. Quarterly Journal of Economics, 75, Feb 1957, 132-150.

Katz, R. L. (1970). Cases and concepts in corporate strategy. Englewood Cliffs: Prentice Hall.

McDougall, P.P. (1987). An analysis of strategy, entry barriers and origin as factors explaining new venture performance. Unpublished dissertation, University of South Carolina.

Miller, A. & Camp, B. (1985). Exploring determinants of success in corporate ventures. Journal of Business Venturing, 1, 87-105.

Sandberg, W.R. (1984). The determinants of new venture performance: strategy, industry structure and entreprenuer. Unpublished dissertation, University of Georgia.

Schoffler, S., Buzzell, R. D. & Heany, D. F. (1974). The impact of Strategic planning on profit performance. Harvard Business Review, 52(2), 81-90.

Skinner, W. (1969). Manufacturing - Missing link in corporate strategy. Harvard Business Review, 47(3), 136-145.

Skinner, W. (1978). Manufacturing in the Corporate Strategy. New York: John Wiley & Sons.

Stocking G. W. &nd Watkins M. W. (1951). Monopoly and Free Enterprise, New York: Twentieth Century Fund.

Von Hippel (1977). Successful and failing internal corporate ventures: An empirical analysis. Industrial Marketing Management, 6, 163-174.

Weiss, C. A. (1981). Start up businesses: a comparison of performances. Sloan Management Review, Fall 1981, 37-53.

Wheelwright, S.C. (1978). Reflecting corporate strategy in manufacturing decisions. Business Horizons. Feb. 1978, 57-66.

15

TECHNICAL AUDIT

How to assess the technological standard of a company

Helmut DETTER (o) Hans H. HINTERHUBER (oo)

(o) Helmut Detter is professor and Head of the Department of Mechatronics, Technical University of Vienna and managing director of ÖCAD, Vienna, Austria.

(oo) Hans H. Hinterhuber is professor and Head of the Department of Management, University of Innsbruck, Austria, and professor of Business Policy, Catholic University of Milan, Italy.

Many top managers and entrepreneurs underestimate the role of technology as a competitive weapon. The sentence that "manufacturing does not matter anymore" is often heard on the management level. This philosophy can endanger the firms competitiveness to a large extent. For this attidude is equivalent to renouncing at strategic thinking and strategic planning.

The aim of this paper is to present a simple procedure for evaluating the technological standard of a company's production system. The evaluation will be made from the production managers and top management's point of view.

Manufacturing as a critical success factor

The decreasing competitiveness of many companies is due to the neglection of the production area. If the sentence "manufacturing does not matter any more" is considered to be valid and strategic thinking is reduced to "management by figures", achieving maintaining susteinable competitive advantages will be very difficult.

The faster the technological progress and the stronger the competition, the more a company needs a technical audit.

The technical audit shows:

1) the actual technical standard of a company relative to competitors,
2) the technical standard required by the strategies at SBU and/or corporate level,
3) ways to get from the actual to the planned standard within a given time and cost limit.

The more detailed the technical audit is, and the sooner it is implemented the greater the chance to achieve and/or maintain a leading market position.

The role of technology in the market life cycle

New technologies undergo a life cycle in three stages (Figure 1):

1) The technical innovation stage:
   One competitor introduces a new production technology. Technical problems and market resistance have to be overcome.

2) Diffusion stage:
   The "first mover" has a competitive advantage. The new technology - a key technology in the meantime - is imitated and quickly adopted by the competitors.

3) Saturation stage:
   The new system is generally used by all competitors. The former key technology becomes a basic technology.

In every stage, different actions must be undertaken to achieve a competitive advantage. The "first mover", by using patents and top security on his technical know how, will try to maintain his competitive advantage relative to his competitors - who still use traditional production techniques.

With increasing experience the "first mover" will be able to produce at a lower cost than the others. He will try to increase his advantage by applying the new technology to other products in his range. The late mover will try to imitate and improve the new system. Or he will try to bring forward his own innovations to restore his competitiveness.

In the saturation stage, when all competitors have access to the new technology, flexibility has to be emphasized. Economies of scope are more important than economies of scale.

The traditional distinction of generic strategies - cost leadership or differentiation - are loosing significance. Barriers of entry for new competitors decrease, whilst competition increases.

Every stage of the market life cycle requires a different Technical Audit.

What does technical audit mean?

Fig. 2 presents a definition of technical audit which emphasizes three main investigation areas.

Each of the three areas has a reference (fig. 3):

a) the competitors, and
b) the international standard of technology.

The comparison with the main competitor and with the international standard of technology indicates opportunities and threats.

The two levels of technical auditing

Two levels of technical auditing have to be considered:

a) the production managers point of view
b) the SBU and/or general managers point of view

Only a dialogue between these two levels can lead to competitive advantages.

a) The production managers point of view:
Operations managers and engineers often only consider the production system as a tool to achieve a high performance and productivity in the most effective way. This attitude is too narrow minded. Nowadays, engineers who just concentrate on technical optimisation, will only fulfil one aspect of their task. Figure 4 shows how the production manager might determine the technical standard of the production system. Different criteria are rated and put into the profile. The sum of relative scores of 11 shows a low standing of the production system compared to that of the main competitors.

b) The SBU and/or general managements point of view:
The SBU and/or general managements attidude towards the production system may be different, considering all relations between the company and the environment or between the SBU's ad their environment. Fig. 5 gives an example of a Technical Audit from the SBU and/or general managements point of view.

The dialogue between production management and SBU or general management, based upon the two audit sheets, indicates:

* The targets and priorities of the actions to be undertaken are jointly set by the SBU and/or general manager and the production manager.
* The methods and ways to attain the targets are set by the production manager himself.

Conclusion

Napoleon is said to have beaten his enemies more with his soldiers' legs than with their bayonnets.

Today companies will gain competitive advantages more with flexible, automated production systems than with traditional production systems. Technical audits brings the technical specialist's and the general manager's perspective together, thus reducing risks and creating new opportunities for sustainable competitive advantages.

Summary:

The paper presents a simple method for evaluating the technological standard of a company's production system from:

a- the technical specialist's

b- the general manager's perspective.

The conclusion is that a Technical Audit is an integrated part of strategy planning; it can make the task easier for a company to be a leading competitor in every market segment in which a SBU operates.

References:

Detter, Helmut und Hans H. Hinterhuber: Anleitung zur technischen Wettbewerbspositionierung. In: "IO Management Zeitschrift" 58 (1989), N.3, 54-58

Hayes, Robert H. und Steven C. Wheelwright: Restoring our Competitive Edge. Compete Through Manufacturing, New York: Wiley 1984

Hinterhuber, Hans H.: Strategische Unternehmungsführung, 4th ed. Berlin-New York: W. de Gruyter 1989

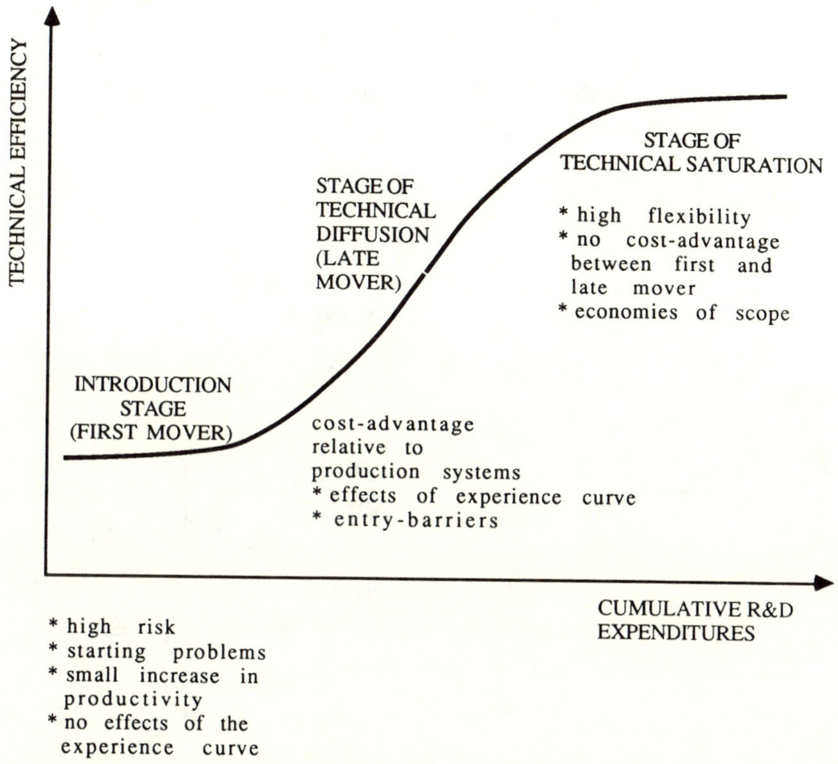

Fig. 1: Stages of development for new production systems

TECHNICAL AUDIT ...

... is a:
* global
* systematic
* periodical
* independent

} evaluation

of how:
* technological developments,
* integration of technological innovation in the strategies at SBU and corporate level,
* management and organizational systems in R&D, logistics, and production

} are analysed and defined,

with the aim of identifying:
* problems, and
* opportunities

concerning:
* actions for creating and/or maintaining sustainable competitive andvantages.

Fig. 2: The meaning of technical audit

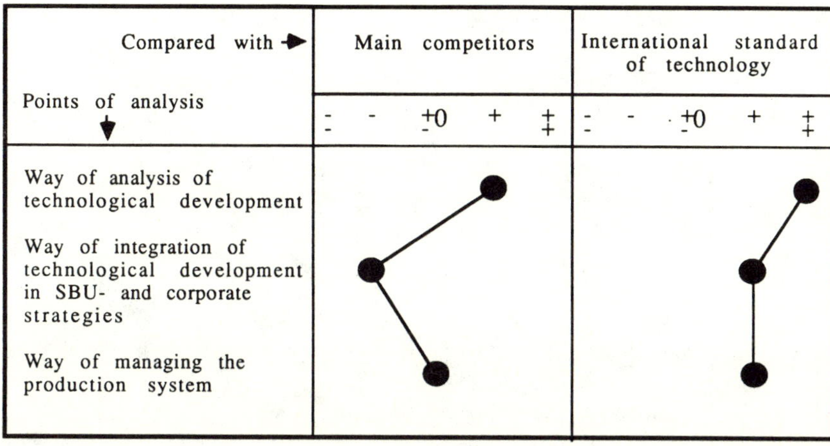

Fig. 3: Reference scheme for Technical Audit (example)

Criteria	Weighting: 3: very important 2: important 1: not important	Evaluation Worse than competitor -1	 Same as competitor 0	 Better than competitor +1	Relative score:	Actions
Input, output, and cycle times	3			●	+3	no action
Operations management	2			●	+2	no action
Capacity/Bottlenecks	2		●		0	remove bottle-necks
Service times	2	●			-2	better service-time by higher flexibility
Ratio direct/indirect employees	1		●		0	rationalisation
Organisation of different production stages	2			●	+2	no action
Systems´flexibility	3			●	+3	no action
Cost - structure	3		●		0	rationalisation
MRP - MPS	3		●		0	improve MRP/ MPS
Failure modes and possibilities	3			●	+3	no action
What´s important? (Suppliers, changing of system etc.)	3	●			-3	Priority 1: improve key problem areas
Quality assurance system	3			●	+3	no action
Sum relative scores (max: +30, min -30):					+11	

Explanation: The higer the relative score is, the better is the competitiveness of the firms production system. A negative score shows an imminent need for improvement, to increase competitiveness.

Fig. 4: Audit sheet for the production manager

Criteria	Weighting: 3: very important 2: important 1: not important	Evaluation			Relative score:	Actions
		Worse than competitor -1	Same as competitor 0	Better than competitor +1		
Uitilisation of machinery	3			●	+3	no action
Flexibility of machinery	2		●		0	increase flexibility
Quality assurance systems	3		●		0	improve quality, planning and control
Dependancy on workers skills and training	2	●			-2	increase training
Stage of automation	3	●			-3	Priority 1: increase automation
Profitability	3			●	+3	increase production
MRP/MPS	2		●		0	no action
Potential for innovation	3		●		0	integrate suppliers
Environmentally friendly	2			●	+2	no action
What can go wrong?	3		●		0	contingency plus prevention planning
Sum relative scores (max: +30, min -30):					+3	

Fig. 5: Audit sheet for SBU or General manager

16

FOCUS, MODERNIZATION, AND MANUFACTURING TECHNOLOGY POLICY@

John E. Ettlie and Joan D. Penner-Hahn

School of Business Administration
The University of Michigan
Ann Arbor, MI 48109-1234
(313) 936-2835

ABSTRACT

In this study we have explored the relationship between plant focus (as measured by strategic emphasis in manufacturing), modernization strategies, and manufacturing technology policy. Based on a study of plants modernizing the production process, the single most important result of these investigations of focus was the finding that manufacturing technology policy was consistently related to focus, regardless of organization size. Firms with aggressive policy have more focused strategically emphasized plants.

We also found significant relationships between delivery focus and new system performance measures such as throughput time reduction and the amount of time to install the new technology. Delivery focus may be easier to implement because it is better understood than flexibility in focus.

Future research directions are suggested, including comparisons of product focus with strategic emphasis focus, broader samples of industries including process flow, and the investigation of flexibility ratios.

INTRODUCTION

Estimates vary but most informed opinions suggest that only about 10% of North American firms have a manufacturing strategy. This excludes firms that have a stated strategy but don't follow it and firms that have not formalized or written down this functional strategy. At a minimum, a manufacturing strategy should answer the question: how will the firm fill orders better than competitors (Hill, 1989). But here we suggest that a manufacturing strategy is actually much more than this. In particular, typical manufacturing strategies need significant enhancement to take into account technological innovation in materials, processes and products.

In 1988, corporate America spent $530 billion on plant and equipment. Caterpillar Tractor alone is in the midst of a $5.9 billion campaign to upgrade the entire corporation. What guides these spending plans? Is there any relationship between manufacturing strategy and how these funds are channelled or the success of these investments?

We review approaches to manufacturing strategy first, and then discuss the importance of technological innovation in these strategies. We report on an empirical study of domestic plant modernization that found manufacturing technology policy to be significantly related to both modernization strategy and focus defined as strategic emphasis. Finally, we propose several research questions from this and other work.

The Literature

There are three widely accepted approaches to characterizing a manufacturing strategy, and a handful of newer views that need to be taken into account. The first approach is Skinner's (1974) concept of focus in manufacturing strategy. Consistency of purpose in manufacturing can be obtained by concentration on product. A variant on this approach is the second method, which redefines focus. This next typology characterizes manufacturing strategy by its strategic emphasis, e.g., cost, quality, flexibility, reliability, and more recently, velocity (Richardson, et al., 1985). The third approach links product and process in a matrix (Hayes and Wheelright (1979)). Variants on this theme are somewhat more complex, but essentially base product-process fit on order size (Hill, 1989). There have been case studies on the manufacturing strategy-making process at Hewlett Packard (Beckman, 1989), at IBM (Bauman, 1988) and Phillips International (Russell, 1989). But this detailed case information has never been incorporated into the content of manufacturing strategy.

Karnani and Brush (1988) found for 429 manufacturing industries in the period 1972-1982 that there was "a move towards plants becoming more focused as reflected in the decrease in vertical integration." The specialization ratio moves only slightly positive during this period, but the authors felt that the latter was a poor proxy for product focus. More recently, evidence has begun to appear suggesting that focus, defined in a more generic way, does promote effectiveness of manufacturing operations. For example, Richardson, et al. (1985) found that corporate focus, defined as "the extent to which a consistent set of parameters in the firm's mission is both selected and given importance relative to other parameters," (p. 20), was significantly related to profit of electronics firms in Canada. Further, they report that "findings suggest that technology-based strategies are most viable," (p. 26), which would be expected in this type of industry.

A recent review of the operations strategy literature, which of course goes beyond functional strategies in manufacturing to include service operations, is representative of the view that operations strategy must be aligned with business or corporate strategy (Anderson, et al., 1989; Hill, 1989). A recent extension of this concept of interdependence is Kotha and Orne's (1989) development of "generic manufacturing strategies." These "generic manufacturing strategies" are linked to Porter's generic strategy framework. With notable exceptions, however, the literature is still dominated by discussions of manufacturing strategy rather than the broader concept of operations strategy.

Even though these extended treatments of operations strategy take into account the role that manufacturing strategy might play in strategic advantage, trade-offs, and distinctive competence, and competition, few have taken technology into account. The view that technology is important in strategy has been largely conceptual to date (Anderson, et al., 1989, p. 15) with little empirical support. We begin to remedy this situation in this report by examining some of the evidence on strategic emphasis focus in modernizing plants. In particular, we hypothesis that both focus and modernization strategy are a function of a cross-functional strategy--manufacturing technology policy--in successful business units.

MANUFACTURING FOCUS AND MODERNIZATION

One aspect of focus that needs clarification is the allowance for new processing technologies now coming on-line in modern manufacturing. These technologies are both more flexible and more integrated with other operations and functions. In this paper we explore the concept of plant focus and the relationship between focus and strategies for modernization as well as the outcomes of these deployment efforts. Hayes and Wheelright (1984, p. 222) say that the development of flexible automation could represent a "third dimension of the (product-process) matrix. This dimension would represent increased overall effectiveness without a major change in the basic match between product life cycle and process life cycle."

Consistent with earlier thinking on focus and recent papers that are germane to this concept (e.g., Skinner, 1986), our thesis is quite simple. Although these newer, flexibly-integrated processing technologies allow greater responsiveness to markets by temporarily focusing on product groups that allow companies potentially greater leverage with components suppliers, they tend to be adopted as part of some overall strategy for business units. Therefore, focus and modernization strategies are expected to be ultimately consistent in the more successful cases. The way in which firms forge the bridge between these two important strategies is of great interest to us because we assume their connection is not automatic. For example, De Meyer, et al. (1987) report that both European and North American manufacturers emphasize quality in their competitive strategies, while the Japanese rate low prices as number one in the Manufacturing Futures Survey. In their action plans, only the Japanese rate flexible manufacturing systems first, while North

American firms are investing in statistical process control and zero defect programs consistent with their strategic emphasis on quality.

We have already reported findings suggesting that a successful modernization strategy is one that matches administrative and technological innovation on degree of radicalness (Ettlie, 1988). It is very likely that some modified concept of implementing strategic focus will enhance the impact of new flexible manufacturing technology on business unit effectiveness. We will test variations of this theme in this paper as well as explore the relationship between flexibility and focus. In particular, we suggest that the paramount bridging or common wellspring concept here that links focus and modernization is manufacturing technology policy.

MANUFACTURING TECHNOLOGY POLICY

Manufacturing technology policy is defined as the long-range plan for modernization of the productive core of an organization and its key interface and support functions like engineering and materials handling or logistics. This definition assumes that in all organizations there is a tendency for gradual change over time to close performance gaps (March and Simon, 1958), and that in manufacturing organizations, some equipment is always in need of replacement because of obsolescence, absence of repair parts and wear-out or breakdown. Therefore, firms vary on the degree to which they aggressively pursue state-of-the-art, radical departures from existing technologies of processing to replace their existing plant capacity in the manufacturing core.

The definition of manufacturing technology policy also needs to be distinguished from several other related concepts. First, it is expected to be distinct from the concept of *innovation strategy*, which in at least one formulation by Freeman (1982, p. 171) focuses on R&D strategy integrated with business unit strategy. Thus, firms can be either offensive, defensive, imitative, etc., in Freeman's terms for their products and services. Obviously, other construct definitions of innovation strategy are possible.

The other distinction that needs to be made is that of technological versus *administrative innovation*. The latter refers to changes in strategies and structures and practices that may or may not be related to technological changes in the operations or design core of an organization. Clearly, an innovation administrative practice would be one that is new to the world, and could apply to manufacturing technology policy, and here the issue of how one measures administrative innovation is important. To the extent that measures of administrative and policy innovation are related, these two concepts may, indeed, be quite similar.

We have spent some considerable effort since 1979 documenting technology policy and the concept of manufacturing technology policy is an important derivative of this work. This earlier work is summarized in Ettlie and Bridges (1987). The more focused version of this concept is discussed in greater detail under the measurements presented below, but the themes have remained relatively consistent over the years and in testing for both published and proprietary samples of over 300 business units. First, there is a tradition and consistency dimension which indicates that a firm does not become strategically innovative overnight. Strategic innovation is something that other people in an industry know about and is a sustained commitment to change. Second, there is a connection with human resource policy. That is, it takes innovative people to be innovative, and firms have to recruit or develop and keep these people fresh to support innovation. Third, innovative policy firms are those that are strongly committed to technological forecasting. Fourth, and finally, firms with an aggressive technology policy advertise their new technology to their customers.

METHODOLOGY

Sample

During the course of a three year study of modernization of domestic manufacturing, we studied a total of 51 plants throughout the United States including one in Canada. These plants were sampled randomly from lists of announcements of plant modernization like those that appear frequently in *Metalworking News* and other trade publications. These plants were visited and data collected in each of the three years. In some cases plants closed, as was the case with some that we were following in the oil well supply industry, and were replaced with comparable installations of flexible automation in plants located in the same state or geographic region of the country. We used interviews and questionnaires and toured each facility where the new processing technology was installed at least once during the course of the study. Data from just the last year (1987) is used in the analyses here when focus questions were included.

Focus

Our measure of manufacturing focus was quite straightforward. We asked the following question: "In your MANUFACTURING STRATEGY, which of the following are most important to you? Please allocate a total of 100 points to the following four items so each score reflects the relative importance of each." Each respondent was handed a list of four strategic foci, with quality first, then cost, delivery and finally flexibility. Many people made comments during the interviews while completing this exercise, and these are discussed later. The score for each plant on manufacturing focus was the standard deviation of these point allocations. The more focus, the more one or two strategies would have to be emphasized, and the greater the standard deviation for the point counts. If all strategies were equally emphasized, with 25 points each, the standard deviation was zero, virtually an unfocused plant. Only one case had zero focus. This method follows Hill and Duke-Woolley (1983) who define strategic focus as "making a consistent and explicit choice of the extent to which aspects such as delivery speed, delivery reliability, quality, cost and product development are provided for by the manufacturing system," (p. 118).

Overall, and in agreement with the De Meyer, et al. (1987), quality was most often emphasized in this sample of domestic plants from our third year data collection. An average of 36.65 points was allocated to quality. Cost was next, with an average of 27.18 points, then delivery with 21.77 points, and finally, flexibility, with 14.55 points. Thirty-four of our 39 third wave plants reported on focus, and one case was eliminated as an outlier (99 points allocated to cost). Average focus for these 33 plants was 13.41 (s.d. = 6.55).

Manufacturing Technology Policy

Manufacturing Technology Policy is the firm or business unit long-range plan for modernization of the productive core. Four, five-choice, Likert-type items resulted from the reliability analysis with a Cronbach alpha of .69, for panel one, an alpha of .46 (n = 34) for panel two data, and an alpha of .72 for third panel data. With the exception of the last item which was new for this study, these questions replicate earlier scales for this construct: 1) a long tradition of being first to try out new equipment and methods; 2) actively engaged in recruiting technical personnel; 3) strongly committed to technological forecasting; and 4) advertise new processing technology to customers. The third panel scale mean was 13.70 (s.d. = 3.81, n = 32), which is slightly above the theoretical mean of 12.0. Although the reliability of the scale is highest in the third panel, it is somewhat lower than other versions we have used, where reliabilities have varied from .88 to .78, but with more items (Ettlie and Bridges, 1987). The decrease in Cronbach alpha for second panel data is assumed to be due to the fine-tuning of this policy during deployment (Ettlie, 1988).

Modernization Strategy

Among other things, we wanted to eventually correlate plant focus scores with several other measures of interest to us in this study. We were, of course, very interested in the relationship between focus and modernization strategy. A plant was scored "1" or synchronous if the degree of radicalness in technology matched the degree of radicalness in administrative innovation adopted to deploy the new manufacturing innovation. For example, in one case, a large robotic assembly system was deployed with maintenance supervision and new software development techniques and was clearly one of a kind. An asynchronous case was scored "0" and represented a mismatch in technology radicalness with administrative approach. Most typical of this type, as would be expected, as adoption of an ambitious new technology in a plant with no new organization structure, no technology agreement with a union, or, more generally, a "business as usual," deployment strategy. Inter-rater reliability for coding on first wave data and three categories of technology and administrative radicalness was $r = .72$ ($p < .01$) with 95% category agreement (Ettlie, 1985). In addition, a panel of six, independent manufacturing engineers was assembled to rank radicalness of first panel technology installations. There was a 77% agreement with the first author on the radical vs. nonradical split in the sample.

Other Measures

There was also a question on the interview schedule that asks whether or not the firm or SBU (strategic business unit) had a written technology policy. This was coded "3" for written, proprietary policy, "2" for written policy and "1" for no written policy. It is assumed that the higher the scores on this scale, the more strategic and valued the technology policy because as coded thusly it was significantly correlated (first panel) with reported "plans in the books" for system investments for the next three years, $r = .25$ ($p < .07$, $n = 39$).

Although validation data are limited, we did find a significant correlation between the percentage of sales spent on computer-integrated manufacturing (average = 2.1%), and scores on the manufacturing technology policy scale, with $r = .77$ ($p = .004$, $n = 10$). There is also a significant correlation between technology policy scores and the reported levels of investment in capital equipment per employee, with $r = .64$ ($p = .013$, $n = 12$), but again, the number of observations is small.

We also included some objective measures of flexibility, like the number of part numbers on the system (average was 570 parts for 23 reporting cases), number of part families, and average change over time in hours (average was 6 hours for 19 reporting cases). A total of eight performance measures for the new manufacturing system were recorded, including achieved cycle time, uptime and utilization. Kendall Tau rank-order correlations were used to validate scores on these three system performance measures with an independent consultant's rankings of overall performance for nine of these plant installations. The resulting taus were .53, .56, and .45, respectively ($.05 < p < .10$). Three indicators of plant size were included: floor space, total number of plant employees and number of blue collar plant employees.

We included self report measures for the impact of the modernization program on 1) improvement in on-time delivery and 2) increased sales or market share. The response format was yes (coded 3), in process (coded 2) and no (coded 1).

RESULTS

Focus and Modernization Strategy

We correlated plant focus scores with deployment strategy, new system performance measures, flexibility scores and plant size indices. These results appear in Table 1. As we expected, there apparently is no simple relationship between plant focus and deployment strategy, as measured by match in technological and administrative innovation. The correlation between plant focus and synchronous strategy was $r = .11$ ($n = 33$, n.s.). The correlation between focus and administrative innovation is also nonsignificant ($r = .11$).

Therefore, we asked the question, under what circumstances are plant focus and deployment strategy related? We examined the thirteen cases that fell above the mean focus score (13.41) for commonalities. There were several. First, and foremost, 11 of the cases were also synchronous, which was a subtle result, not detected in the correlation analysis. Of the 22 cases below the mean focus score, only 12 were synchronous.

The cases that were above the mean focus score and also synchronous in deployment strategy had some similarities. They were either small plants (four of the 11 had 200 or fewer employees on site), or were remarkable in the degree to which the actual installed system was integrated with the rest of the plant. This is exactly what one would expect when searching for reasons why a modernization and plant focus strategy would be consistent. Another thing these 11

Table 1: Correlation Between Focus and Performance Outcomes

	Synchronous	Parts	Families	Change-over	Cycle Time	Uptime	Utilization	Unanticipated Expenses	Scrap & Rework	Throughput Time Reduction	Time to Install	Floor Space Reduction	# of Plant Employees	# of Plant Bluecollar	Plant Floor Spaces
focus	.1127 (33) p = .266	-.2420 (23) p = .133	-.1374 (20) p = .282	-.2147 (19) p = .189	-.1241 (28) p = .265	-.2712 (31) p = .070	-.2601 (30) p = .083	.1237 (26) p = .274	.1618 (23) p = .222	.0119 (22) p = .479	.3328 (25) p = .052	-.1447 (23) p = .255	-.0541 (33) p = .382	.0118 (32) p = .474	-.0479 (30) p = .401
quality	.0099 (33) p = .478	-.1106 (23) p = .308	.0012 (20) p = .498	-.1815 (19) p = .229	.0356 (28) p = .429	-.2575 (31) p = .081	-.0436 (30) p = .409	.0147 (26) p = .472	.0922 (23) p = .338	-.3129 (22) p = .078	.3330 (25) p = .052	-.0699 (23) p = .376	-.0366 (33) p = .420	-.0046 (32) p = .490	.0270 (30) p=.444
cost	.1831 (33) p = .154	-.0818 (23) p = .355	.1040 (20) p = .331	-.1243 (19) p = .306	.1252 (28) p = .263	.2498 (31) p = .088	-.0254 (30) p = .447	.0574 (26) p = .390	-.1815 (23) p = .204	.1582 (22) p = .241	.1519 (25) p = .234	-.1323 (23) p = .274	.2893 (33) p = .051	.2763 (32) p = .063	.2790 (30) p = .068
delivery	.0976 (33) p = .295	.0668 (23) p = .381	-.3217 (20) p = .083	.1882 (19) p = .220	-.0447 (28) p = .411	-.1272 (31) p = .248	-.1688 (30) p = .186	.2030 (26) p = .160	.2953 (23) p = .086	.3808 (22) p = .040	-.4017 (25) p = .023	.1663 (23) p = .224	-.1549 (33) p = .195	-.1086 (32) p = .277	-.2703 (30) p = .074
flexibility	-.3554 (33) p = .021	.2092 (23) p = .169	.2518 (20) p = .142	.1618 (19) p = .254	-.1495 (28) p = .224	.1435 (31) p = .221	.2772 (30) p = .069	-.2774 (26) p = .085	-.2465 (23) p = .128	-.2284 (22) p = .153	-.2687 (25) p = .097	.0632 (23) p = .387	-.1471 (33) p = .207	-.2225 (32) p = .110	-.1003 (30) p = .299

plants had in common is the remarkable flexibility of operations. One of the most outstanding cases of flexible integration fell within this group. It is an assembly cell for an automotive component and supplies, with almost instantaneous changeover, a regional assembly plant on a four-hour just-in-time delivery system.

Note that flexibility, per se, as measured by number of parts and part families as well as change over times, was not significantly correlated with focus or deployment strategy. Thus, flexible integration appears to be one of the paramount contingencies that has to be manipulated to forge the link between focus and modernization strategies.

The two cases that did not follow this pattern are also worth examining. One was a corporate technology experiment in a plant. It resulted in high R&D returns but low productivity returns to the business unit. The second case, also a low performer, was in a new plant with many start-up problems, including human resource difficulties. Focus in this latter plant has changed several times after it was designed and built.

We found no significant relationships between the total measures of administrative innovation (hierarchy, design-manufacturing, suppliers, and customers) and focus. Finally, we found two statistically significant relationships for focus and outcomes for modernization. Focus was significantly correlated with return on investment for the project, with $r = .53$, $p = .025$, $n = 14$). Focus was also significantly correlated with larger labor savings on the modernization project with $r = .61$ ($p = .002$, $n = 20$).

Focus and Modernization Impact on Customers

We asked respondents if their modernization program had any impact on improving on-time delivery to customers and if sales or market share had improved when the new technology manufacturing system was in place. The correlations between plant focus and this modernization impact--which allows temporary focus on product groups--were both nonsignificant. For sales improvement, the coefficient was in the predicted direction, with $r = .21$ ($n = 16$, n.s.), but not large enough to be beyond the chance level. Cost focus did have an inverse relationship with improved on-time delivery with $r = .38$ ($p = .027$, $n = 26$).

Although focus, as measured here, did not seem to be related broadly to modernization outcomes affecting customers directly, these same customer impact reports were correlated with other factors. For example: reports of modernization increasing sales or market share were significantly and inversely correlated with grievances (yes = 1, no = 0), with $r = -.47$ ($p = .083$, $n = 15$) and approached significant correlation with customer administrative innovations with $r = .40$ ($p = .090$, $n = 12$). Improved on-time delivery was significantly correlated with supplier integrating innovations, with $r = .57$ ($p = .018$, $n = 14$). Again, these results suggest that plant focus, per se, does not necessarily have a simple relationship with modernization, and modernization impact on customer, from the user's perspective.

Quality Focus and Modernization

Based on the correlation results of Table 4, there does not appear to be any simple relationship between modernization strategy, as measured here and the typical priority focus in domestic plants, quality. However, it should be remembered that this sample of firms was selected because they are embarking on significant modernization. If quality is typical of their focus and they are also modernizing, there should be a relationship between quality and modernization. There is.

Scrap and rework rates are significantly lower for the new technology systems at an average rate of 2.91% of total manufacturing cost versus 4.26% in existing plant. The t-statistic is significant at $t = 2.78$ ($n = 23$, $p = .011$).

Delivery and Modernization

Other results in Table 1 indicate that there may be additional connections between modernization strategy and plant focus, perhaps springing from the underlying dimensions of business unit strategy. There was a significant correlation between delivery focus and throughput time reduction achieved by the new system, $r = .38$ ($p < .04$). Further, there was a significant inverse relationship between delivery focus and time to install the system, with $r = -.40$ ($p < .02$). It seems clear that plants that are emphasizing delivery in their focused strategy have been successful at using new technology to implement this strategy. This is the only clear, consistent relationship we were able to detect, with these analyses, between the actual substance of strategic focus and modernization outcomes.

Flexibility in Focus and Modernization Strategy

Somewhat unexpectedly, we found that there was a significant, inverse relationship between flexibility in strategic focus and modernization strategy, with $r = -.36$ ($p < .02$). That is, plants that reportedly were more focused on flexibility in their strategy, were less likely to match their technological and administrative innovations for deployment. We expected there would be no relationship to a mildly positive relationship between these two indicators, because flexibility without being integrated into the plant, actually detracts from focus, and therefore would drive a wedge between these various means of implementing business unit strategy. So there are a number of reasons why this result may obtain.

1. It may be that plants that emphasize flexibility as an end in itself in the domestic plants have not as yet learned how to capitalize on this new capability in productive systems. We say this because synchronous plants have already been found to have significantly higher uptime, cycle times and utilization, as well as being faster to install new, flexible systems.

2. Plants that focus on flexibility may have had only marginal success in actually achieving flexibility, and, therefore continue to pursue it up a learning curve. Note that, although all the correlations between flexibility focus and objective measures of flexibility (parts, part families and change over) are positive, none is statistically significant.

3. Third, flexibility is so rare as a strategic focus and emphasis in domestic plants, in spite of the widespread adoption of flexible automation, that it is quite possible that we are seeing the emergence of a new, quite different strategy for competition that is still not well understood. Since only the Japanese are apparently now adopting flexibility as a priority manufacturing strategy, based on the De Meyer, et al. (1987) findings, it may be quite difficult and premature to interpret these findings and this trend. Our other indications are that flexibility must be integrated in order to be an effective strategy. The two cases that scored highest on flexibility and that are near the mean on focus are also high volume systems. One of these cases, a high volume FMS implements a unique strategy of gaining leverage with suppliers. The strategy is to shift production from suppliers to this unique high volume, flexible facility in order to gain a cost advantage and prevent suppliers from capturing excess profits. This is quite unusual in a world where most flexible systems are being pursued for market reasons. In another case, flexibility was used to play the metal commodity markets by changing material inputs.

4. Fourth, and finally, flexibility is a means to an end, not an end itself. If used wisely, it can support any focused strategy. However, there are difficult types of flexibility to achieve and easier flexibility goals. Part family additions are assumed to be harder than part numbers. Therefore, the *ratio* of families to parts should be considered a research variable.

Focus and Manufacturing Technology Policy

We hypothesized that the bridging or common concept for focus and modernization would be manufacturing technology policy, and the results indicate that it is. Firms with an aggressive technology policy have more focused plants. The correlation between focus and third panel manufacturing technology policy is $r = .36$ ($p = .056$, $n = 21$). Although this correlation coefficient just fails to be significant at traditional cut-off levels, we are confident, based on earlier work and, based on the result that administrative innovations scores that go into determining synchronous versus asynchronous status of these cases were also directly related with technology policy, with $r = .31$ ($p = .066$, $n = 25$) that this policy is the key to understanding focus and modernization. In the first panel of data administrative innovation and technology policy were significantly correlated only for cases where the level of investment was more than $3,000 per plant employee. Technology policy and the individual dimensions of focus (i.e., quality, cost, delivery, and flexibility) were not significantly correlated. Although the number of complete data cases is small, technology policy was significantly correlated with the number of parts families, r = .49 (n = 13, p = .014) and just failed to be significantly related to the number of parts scheduled on new systems, r = .40 (p = .052, n = .7).

Since all were related to organization size, we controlled for size with partial correlation and the result was persistent. The correlation between focus and manufacturing technology policy, controlling for organization size, was r = .43 (df = 7, p = .126). Again, sample size is small, so one has to be cautious about drawing conclusions, but the consistency is there. Focus and administrative innovation were not significantly correlated (r = .12, n = 33, n.s.). Recall that focus was significantly correlated with return on investment for the project, with r = .53, p = .025, n = 14). Focus was also significantly correlated with larger labor savings on the modernization project with r = .61 (p = .002, n = 20). These relationships are summarized in Table 2.

Table 2: Summary of Correlation Results in Proposed Causal Form

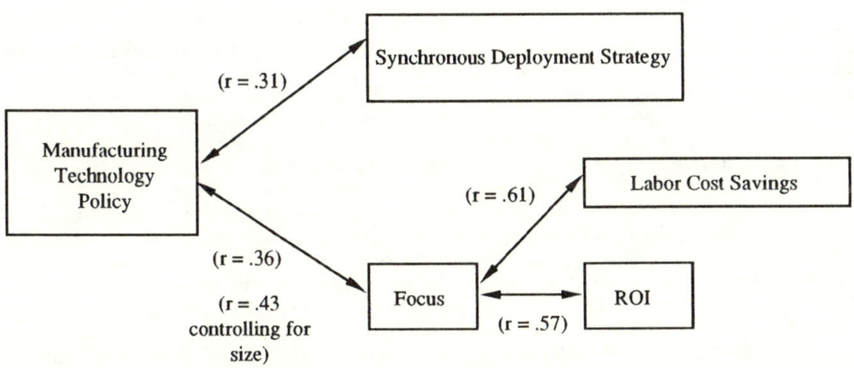

We believe there is great hope for further understanding of the focus, modernization, technology policy relationship with more development of concepts and testing detailed hypothesis in other sectors of the economy. If focus and modernization approach both spring from the same well, approximated by manufacturing technology policy here, then the links of causation should be there to be discovered by a causal study.

RESEARCH QUESTIONS

Focus as Strategic Emphasis

The results from our study indicate that when focus is defined and measured as strategic emphasis it is likely to be associated with manufacturing technology policy in discrete parts industries. We have shown elsewhere (Ettlie, 1983; Ettlie, et al., 1984) that technology policy is consistently and significantly correlated with business unit strategies that emphasize new product launches. This appears to be one empirical verification of the often hypothesized connection between business strategy and manufacturing strategy. Now it also appears that a more aggressive manufacturing policy is likely to be correlated with greater focus in plants as defined by strategic emphasis.

Several questions still remain. We have not ventured outside of batch manufacturing, assuming one includes food processing in batch (Ettlie, et al., 1984), so a more comprehensive test of the technology policy focus hypothesis is needed. What is more, flow-type processing typically has batch manufacturing or batch packaging and logistics associated with it, so a mix is typical. Theoretical refinements are obviously in order to account for these complex operations and then further empirical studies will be warranted.

What is more, all these approaches suggest contingency notions of manufacturing strategy, and it cannot be assumed that this is always appropriate. For example, earlier we found, among other things, that firms in batch manufacturing that clearly understood group technology before they adopted some type of flexible automation were much more likely to take a calculated risk approach toward modernization which included using some type of administrative innovation to deploy new advanced manufacturing technology in their plants (Ettlie, 1988). It may be that in batch manufacturing, understanding group technology is not only necessary but essential for success not only in manufacturing but in technology sharing, as well (Rubenstein, 1989). We still know little about the relationship between focus as strategic emphasis and focus as product or product group emphasis. This should become a key research question.

Focus on Product

Manufacturing capacity that is focused on products and customers has inherent advantages, it would appear. But our experience tells us that there are three important exceptions to this principle that need investigation.

First, it is possible that a product or customer focused strategy only works for manufacturing capacity that can be physically isolated from the rest of the plant. The reason for this statements is that we are aware of two rather significant failures of the product focused factory concept and they have several common traits. For one, both plants split into four or five focused operating units, pretending to be "800 miles apart" in one case. It just doesn't work. The union and white collar workers wanted opportunities for advancement which they can't exploit under this strategy. As one begins to fine tune the product focused strategy, it sometimes becomes more watered down. Close proximity to suppliers becomes more important for unit consolidation and those other complications set in. In one of these cases, the union struck over issues that included focused factory. Outsourcing is not a trivial issue.

Second, flexibility in manufacturing has not only jeopardized a pure focused strategy because it allows customer shifting, at least one case we are familiar with uses flexibility to manage suppliers as well. This firm is experimenting with the concept that suppliers with big margins need to kept honest and going from one component supplier to the next is a new way of managing the make-buy decision. That is, vertical integration is really part of manufacturing strategy as well as business strategy.

Third, we are aware of a company in West Germany that adopted flexible manufacturing as well as a strategy to "make heavy parts and source the rest" in the paper processing equipment industry. Manufacturing strategies can be simple! They often are insensitive to component type when flexibility in technology and people is also introduced as a strategy.

We suggest that the concept of *flexibility ratios* be adopted to rescue some of these issues so that flexibility can effectively be introduced into this debate. We are pursuing this approach.

REFERENCES

Anderson, John Cleveland, Gary, and Schroeder, Roger G., "Operations Strategy: A Literature Review," *Journal of Operations Management*, Vol. 8, No. 2, Spring 1989, pp. 1-26.

Baumann, L. Joseph, "IBM Manufacturing Strategies for Competitiveness" presented at the Second Annual Manufacturing Forum, University of Michigan, Ann Arbor, Michigan, November 22, 1988.

Beckman, Sara L., "Exercising Breakthrough Thinking in Manufacturing: A Hewlett-Packard Case Study" presented at the College on Innovation Management and Entrepreneurship, The Institute of Management Sciences, Vancouver, Canada, May 8-10, 1989.

De Meyer, Arnoud, Nakane, Jinichiro, and Miller, Jeffery G., "Flexibility: The Next Competitive Battle," working paper No. 86/31, revised March 1987, INSEAD, Fontainebleau France.

Ettlie, J. E., *Taking Charge of Manufacturing*, San Francisco, Jossey-Bass Publishers, 1988.

Ettlie, J. E., The R&D Management Literature and Innovations Diffusion Research," with Radnor, M. and Dutton, J., Chapter 5 in the *Diffusion of Innovation: An Assessment*, edited by M. Radnor, I. Feller, and E. Rogers, Center for the Interdisciplinary Study of Science and Technology, Northwestern University, Evanston, IL, July 1978.

Ettlie, J. E., "Organizational Adaptations for Radical Process Innovations," presented at the Symposium Organizational Research on Advanced Manufacturing Technology, at the 45th Annual National Meeting of the Academy of Management, San Diego, CA, August 11-14, 1985.

Ettlie, J. E., Organizational Policy and Innovation Among Suppliers to the Food Processing Sector," *Academy of Management Journal* 26, 1, 1983.

Ettlie, J. E. and Bridges, W. P., "Technology Policy and Innovation in Organizations." in J. Pennings and A. Buitendam (eds.), *Technology as Organizational Innovation*. Cambridge, Mass.: Ballinger, 1987.

Ettlie, J. E., Bridges, W. P., and O'Keefe, R., "Organizational Strategy and Structural Differences for Radical versus Incremental Innovation," *Management Science*, Vol. 30, No. 6, June 1984.

Freeman, C., *The Economics of Industrial Innovation*, 2nd Ed., Cambridge, Mass.: MIT Press, 1982.

Hayes, Robert H., and Wheelright, Steven C., *Restoring Our Competitive Edge*, New York, Wiley, 1984.

Hayes, Robert H., and Wheelright, Steven C., "Link Manufacturing Process and Product Life Cycles," *Harvard Business Review*, Vol. 57, No. 1, January-February, 1979, pp. 133-140.

Hill, Terence J., and Duke-Woolley, R. M. G., "Progression or Regression in Facilities Focus," *Strategic Management Journal*, Vol. 4, 1983, 109-121.

Hill, Terry, *Manufacturing Strategy*, Text and Cases, Homewood, IL, Irwin, 1989.

Karnani, Aneel, and Brush, Thomas, "Trends in Size of U.S. Manufacturing Plants: An Empirical Study,": Working paper, School of Business Administration, The University of Michigan, 1988.

March, James G., and Simon, Herbert A., *Organizations*, New York, NY, John Wiley & sons, 1958.

Richardson, P. R., Taylor, A. J., and Gordon, J. R. M., "A Strategic Approach to Evaluating Manufacturing Performance," *Interfaces*, Vol. 15, No. 6, November-December, 1985, pp. 15-27.

Russell, Kathryn, "The Consumer Electronics Turnaround Lean and Mean," Mission Critical: Strategic Manufacturing Systems, Yankee Corp, Conference, Chicago, In, April 11-12, 1989.

Rubenstein, Albert H., *Managing Technology in the Decentralized Firm*, New York, NY, John Wiley & Sons, 1989.

Skinner, Wickham, "The Productivity Paradox," *Harvard Business Review*, Vol. 64, No. 4, July-August, 1986, pp. 55-59.

Skinner, Wickham, "The Focused Factory," *Harvard Business Review*, Vol. 52, No. 3, May-June 1974, pp. 113-121.

Van de Ven, A. "Central Problems in Management of Innovation," *Management Science*, Vol. 32, No. 5, May 1986.

@Work in this area was supported in part by the National Science Foundation, but the authors are solely responsible for content.

17

ILLUMINATING THE BLIND SPOT: EXAMINING LINKAGES BETWEEN MANUFACTURING TECHNOLOGY AND COMPETITIVE STRATEGY

Dean M. Schroeder, Steven W. Congden, and C. Gopinath

School of Management
University of Massachusetts
Amherst, Massachusetts 01003
(413) 549-4930

ABSTRACT

This paper presents an empirically grounded study investigating linkages between manufacturing technology and competitive strategy in light of fundamental changes in technology due to flexible, computer controlled equipment. The technology - strategy relationship and the alignment process is examined for twenty firms. The purpose of the paper is to raise questions, rather than answer them, and identify promising directions for manufacturing strategy research.

INTRODUCTION

Not since the Industrial Revolution has manufacturing experienced such radical change in technologies. The rigid, mechanical equipment of the Industrial Age is being replaced with flexible, computer controlled technologies from the Information Age. Manufacturing strategy and process technologies are increasingly being recognized for their strategic importance as Japanese firms gain leadership in many industries by closely integrating manufacturing technologies into their competitive strategies (Buffa, 1984; Jaikumar, 1986). Unfortunately, U.S. manufacturers have been less successful than foreign competitors in taking advantage of the new smart technologies (Dertouzos et al., 1989; Jaikumar, 1986; Piore and Sabel, 1984). Consequently, the blind spot in strategy that Skinner (1969) drew to our attention two decades ago has become a critical focus.

This study identifies and examines issues related to aligning a firm's manufacturing equipment technology with its competitive strategy given the advent of computer controlled technologies. Twenty manufacturers are examined and compared to identify technology-strategy linkage patterns. Rather than test, the purpose is to explore the impact of new flexible equipment on technology-strategy alignment patterns and the process by which these alignments emerge.

LITERATURE

The traditional view of process technology's strategic role involves a trade-off between flexibility and efficiency. A stream of research initiated by Abernathy (e.g., Abernathy, 1976; Abernathy & Townsend, 1975; Abernathy & Utterback, 1978, Hayes & Wheelwright, 1979) found this trade-off to evolve predictably over a product's life cycle. Initially, when competition centers around product innovation, flexible, general purpose processes are required to accommodate product variations and frequent design changes. As the product standardizes and volume rises, competitive concerns shift to price and production systems become more dedicated (less flexible) in an effort to lower costs. This dynamic culminates in complex, integrated, rigid, and capital intensive processes.

Although this evolution of processes works to the advantage of those with large market share (Utterback, 1979) and increases capital entry barriers, new process technology can alter this relationship. Schroeder (1989) found that a manufacturing innovation in the Foundry industry allowed smaller job-shops to effectively compete on a cost basis in markets previously the exclusive domain of large, dedicated producers. In this manner, Schumpeter's (1934) force of "creative destruction" threatens the status quo of competitive patterns.

Computer controlled process technologies represent such a force for many industries and have prompted scholars to speculate that the traditional flexibility-efficiency trade-off may be radically altered (Blois, 1985; De Meyer et al., 1989; Jelinek & Goldhar, 1983; Hayes & Wheelwright, 1984; Meredith, 1987; Thompson and Paris, 1982; Voss, 1986). Computer controlled processes are argued to dramatically lower the cost of flexibility, making it nearly as efficient to manufacture product variations as it is to manufacture large volumes of standard products. Jelinek & Goldhar (1983) assert that flexible technologies are replacing price competition on commodity-like products with competition based on special options, custom products, and uniqueness.

That manufacturing strategy involves more than a flexibility - efficiency trade-off was clearly recognized prior to the advent of computer controlled technologies. Skinner's (1974) "Focused Factory", in which manufacturing is organized around the needs of product/market strategies, recognized competitive dimensions of low cost, product quality, dependable delivery promises, short delivery cycles, flexibility to produce new products quickly, flexibility in adjusting to volume changes, and low investment. With variation in terms and categories, Hayes & Schmenner (1978), Stobaugh and Telesio (1983), Wheelwright (1984), and Swamidass (1987) present similar lists.

Inherent in this expanded view is the idea that a given process technology may only be suitable for certain strategies. If equipment and/or procedures are tailored to specific dimensions of strategy, it stands to reason that they would be less than optimal with respect to other strategic dimensions. Trade-offs in policies and equipment make it difficult for a factory to perform well on every dimension (Skinner, 1974, 1985; Hill & Duke-Woolley, 1983).

While the inability of given process equipment to serve every strategic dimension is well recognized, research has yet to determine whether a good "fit" between strategy and process technology results in higher performance (Kotha & Orne, 1989). Furthermore, what has been learned must be re-examined in light of the significant impact of computer controlled technologies. This study examines the linkage patterns among both firms with traditional and firms with computer controlled manufacturing technologies and their competitive strategies. Our goal is to explore the nature of technology-strategy fit and identify promising directions for future research.

METHODS

Consistent with the exploratory nature of this study, a grounded theory approach (Glaser & Strauss, 1967) was utilized in which theory is discovered from observational data in a systematic manner. A multiple case design focusing on twenty manufacturers in three industries was employed to control variance which complicates more global studies yet provide sufficient diversity to illuminate different process technology-strategy linkage patterns.

The sample consisted of eight machine job-shops, four metal cutting tool manufacturers, and eight plastic injection molding firms. These industries were selected because of their highly competitive environments, recent introduction of new computer controlled process technologies, and importance to industrial economies. Small to medium firms (14 - 140 employees) were selected because their operations, by nature, more tightly link manufacturing and competitive strategies.

An effort was made to select a stratified sample to include both highly profitable and less profitable firms, and firms using advanced manufacturing technologies and those using more traditional technologies. However, the ultimate sample was not perfectly balanced as firm categorization was based on research findings rather than a priori information. The measures used to determine these categories were developed from the research and are discussed later.

Data gathering began with background industry and technology studies designed to identify competitive forces and critical technologies. The field research involved interviews, plant tours, and questionnaires. Key mangers (usually the CEO) were interviewed using a semi-structured format covering both general and specific manufacturing, competition, strategy, and technology issues. Interviews, averaging 90 minutes, were recorded and transcribed into 520 single spaced typed pages of textual data. Additional observations recorded during plant tours confirmed the use of various technologies. Research questionnaires were distributed to augment interview and observational data.

The data were analyzed by each member of the research team and compared. Interpretations were cross checked with questionnaire responses and, whenever possible, direct quotes from the interviews.

The variables of primary interest to this study were process technologies used, strategy followed, and financial performance. The technologies of principle focus were computer numerically controlled (CNC) machine tools in the metal machining industries and computer process controls in the plastic injection molding industry. Firms were classified as low technology if they created less than 10 percent of their manufacturing value with computer controlled technologies, medium if between 10 and 50 percent, and high if more than 50 percent. These groupings offered relatively clear demarcations among the firms.

Firm performance is always tricky to measure. Deficiencies of accounting measures are well documented (e.g. Fisher & McGowan, 1983; Salamon, 1985; Chakravarthy, 1986). This is especially true in small firms where variations in owners' compensation can cause relatively major distortions of profit margins. In addition, such firms are typically privately held and reluctant to reveal performance numbers. Consequently, two techniques were used to measure performance. First, managers were asked about profitability and growth. This initiated discussion designed to gain the fullest possible understanding of performance without alienating the interviewee. Those managers willing to reveal full profit and loss numbers were rare, but all did provide good general indications of performance. High performers were growing in typically stagnant markets and were relatively more profitable than competitors. Low performers were, at best, breaking even. Second, the questionnaire asked managers to classify their firms' performance relative to competitors on a 5 point scale similar to one found by Dess and Robinson (1984) to be highly reliable. Results from the two measures were compared with high correlation.

To measure competition and strategy, managers were asked both open ended and specific questions related to the manner in which their firms competed. In keeping with the nature of the study, dimensions of strategy that meaningfully relate to manufacturing technology were allowed to emerge without a priori conjecture. These were then compared with existing strategy typologies, including variations of Porter's (1980) generic competitive strategies and strategic dimensions proposed by Skinner (1974) as relevant to manufacturing.

OBSERVATIONS

This section begins by examining observations on competitive strategies and technology usage patterns, followed by exploration of patterns in market position/product mix strategies. Focus then shifts to the process by which technology - strategy alignment is achieved.

Figure 1 provides an overview of our sample firms in terms of their performance and level of computer controlled (CC) technology utilization. A coding system which numbers and categorized companies by industry (i.e. M for job-shop machining, T for metal cutting tool maker, and P for plastic injection molder) is used both in the figures and throughout the paper as a reference to the specific firms.

Strategies

The competitive dimensions most commonly reported were price, quality, and service. Service was in terms of delivery, design help, value added, etc. When viewing these from the perspective of Porter's generic strategies, all the firms had at least a minimum level of differentiation and customer focus, typically a market niche being served which required specific capabilities. Within these niches, however, all competitors possessed similar capabilities. Consequently, in most cases, price played the dominant competitive role. Even in the highest quality shops price was critical:

> "Number one, I guess, is capability - can you make the part. Number 2 would be price and delivery and the price better be damn competitive."(T1)

> "Quality is well advertised, but you'd better have the lowest price."(M1)

This is not saying there were no advantages to being able to offer better quality, service, or delivery. "Price is still a major factor. But not the only factor."(P5) Once a firm's price matched that of the lowest priced competitor, other factors came into play:

> "In our business its the service that's different and the customer is going to demand the price."(P2)

The dominance of price competition among all rivals with a given set of capabilities appears to hold for all markets segments except for shops producing unique prototype and complex, custom, one-of-a-kind work. The nature of this type of work dictated reputation and trust as more important than price:

> "In this kind of work I think the service end of it is super important because you're working with a product which we sometimes have to take latitude with and understand what it is our customers are trying to accomplish. Knowing what they do is very important." (M4)

These firms were producing work which had no market in the classic sense. The products were customer specific, lacked multiple transactions to set market prices, and involved high outcome uncertainty. Thus normal market mechanisms fail (Williamson, 1975) and trust and close customer relations became more important that price.

No direct patterns between the use of CC technology and generic competitive strategies were observed. This was largely because competitors for a given type of work all had similar capabilities and provided like quality, precision, etc. Being all relatively equal, price competition dominated. However, when comparing technology use across groups, a different picture emerges. There were several clear patterns linking technology use and the market segments within which firms competed. The dimensions which most clearly differentiated market segments in relation to technology were: 1) batch size, 2) level of precision, and 3) degree of contract manufacturing.

Batch Size

Firms producing large batch sizes and/or repeat orders of parts tended to perform better when using more CC technologies than firms that produced small lot-sizes or one-of-a-kind products. This pattern is highly visible in Figure 1. Of the higher performing firms (left side of the Figure 1) those with larger batch sizes clearly used more CC technology than those with small batches. This relationship reverses for the low performing firms on the right side of Figure 1.

The strength of this "fit" between batch size and the use of computer controlled equipment is most observable when examining firms that had an apparent mismatch between batch size and technology. One manager/owner expressed the dilemma quite clearly:

> "The whole problem is we're marketing for the guys that are in a onesy, twosy business. And we've got these [CNC] production machines. So we're always strangers in the production world... We get more utilization out of our manual machines." (M7)

FIGURE 1
PRODUCTION BATCH SIZE

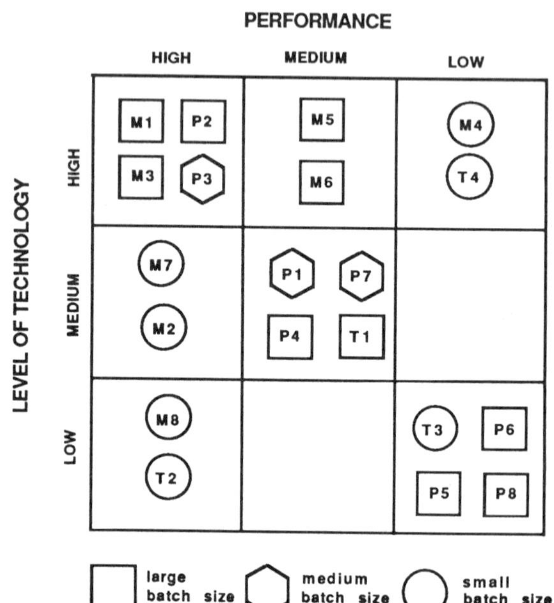

M2, a shop that specialized in one-of-a-kind and custom small batch prototype work, experienced strategy fit problems after adopting CNC equipment. They accepted a three year (minimum) high volume contract from a major defense contractor and purchased CNC machine tools for the work. One year into the contract the government cancelled the program. M2, left with a poor fit between its process technology and work, altered its strategies and began pursuing larger batch work in a effort to utilize its idle CC equipment.

Similar misalignments were experienced by T4 and M4. At T4 the accountant, sales manager, and an equipment salesman convinced the owner that state-of-the-art CNC grinding equipment was perfect for the firm's work, which was one-of-a-kind and small batch metal cutting tool resharpening, tool customizing, and special tool manufacturing. Once installed, the equipment was found to lack the required flexibility and was much more time consuming to program than anticipated. Because the equipment represented a non-recoverable, "bet-the-company" investment, T4 was attempting to totally shift their strategic focus from a tool sharpener and customizer to a volume tool manufacturer.

M4's specialty was one-of-a-kind rubber injection molds requiring extensive design engineering. The company principals' strong technical interests lead to habitual adoption of the latest production equipment. Unfortunately, while the firm's engineering work was complex, the production work could easily be done using conventional machine tools and could not justify the overhead burden and programming costs of new CNC equipment. M4 had not made a profit in 5 years.

With the evolution of CC technologies toward increasing flexibility and user friendliness, their use was observed to be increasingly appropriate for smaller and smaller batch work. As M5 explained:

> "CNC....was best employed on high quantity because it was so difficult to get it set up and adjusted properly. Now that is changing... We're getting to the point where now we can do relatively simple parts in small quantities (as low a 50 units) and do it competitively with a manual setup."

This move toward smaller lot sizes appears to be most prevalent with firms which require the extra precision or other performance characteristics CC equipment offers. In some cases, the work being done is simply not possible without such technology.

Precision

This leads to the second strategic market position for which CC technologies were important, high precision parts. By high precision we mean the ability to maintain close tolerances and repeatable results.

For many shops serving customers requiring greater precision, use of advanced technologies was simply an issue of capability: without such advanced equipment the precision jobs were difficult or impossible to produce. With CC technologies "you can hold tolerances that before you just couldn't even think [of]" (President M6). In plastic injection molding, computerized process controls recorded molding parameters for customers requiring such data for quality control or R&D. As observed in Figure 2, the plastics firms competing for customers with highly sophisticated end products, P2, P3, and P4, all employed the latest CC process technologies. The customers were in the aerospace, computer, photographic equipment, or medical instrument industries where product applications demanded the highest levels of quality and the tightest tolerances. This was clearly expressed by a P4 manager:

> "It [adoption of the latest technology] was driven by our market...they are the leading companies in the world and they have to stay on the edge of innovation and we therefore have to stay there too."

Figure 2 presents a somewhat inconsistent picture, however, particularly in the case of machine shops. For example, M2, M7, and M8, who are marketing to customers with high precision needs, are ranked only low or medium in technology. This is primarily because the technology ranking measure used focuses on major pieces of CC manufacturing equipment. While this provides a highly visible and documentable decision rule, it misses many subtle technology differences. The small lot firms with high precision requirements could not justify widespread use of major CNC equipment because they lacked sufficient repeatable volume. Yet they did employ minor CC technologies and other advanced manufacturing equipment more appropriate for their needs, such as an advanced cutting fluid filtering system (T2 & M8), sophisticated dimensional measurement tools (M8 & M1), glass scales with digital displays to increase accuracy on conventional machine tools (M8), and CNC electronic discharge machines (M8). The critical issue appears to be the choice of appropriate advanced technology to fit the firm's strategic position, not simply having CNC equipment. Advanced manufacturing technologies are required for the capability provided in serving customers. The particular technology which is appropriate depends on the specific needs being served.

FIGURE 2
MARKETS SERVED

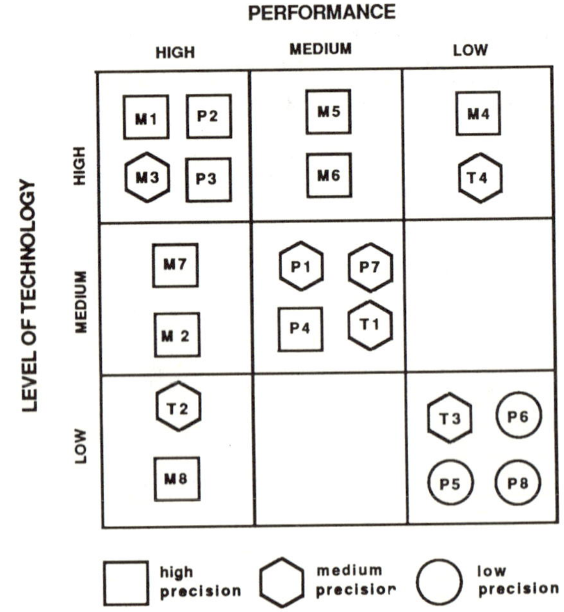

Level of Contracting

A third and surprising pattern was observed when contrasting contract manufacturers with firms producing proprietary (their own) products. Firms with proprietary products used less CC technology than contract shops (see Figure 3). With the link between large batch size and more CC technology, firms with proprietary products were expected to benefit more from such technologies because of longer production runs and more stable product mixes. Yet the opposite was observed. A follow-up study (Congden & Schroeder, 1989, - available from the authors) of 183 firms confirmed this relationship.

We have speculated on three explanations for proprietary firms using less CC technology. First, firms with their own products have more places to gain competitive advantage and are thus less narrowly focussed on manufacturing. Second, contracting firms sell a service which is less tangible to evaluate than a product; therefore, having the latest technology can be important in projecting a capable image. In a study of CNC usage by small U.K firms Dodgeson (1985) found contracting firms believed that CC technology impressed potential customers. Third, firms producing proprietary products are in less fragmented markets than contracting firms, and thus may face less competitive pressure to adopt because of oligopolistic market behavior.

THE ALIGNMENT PROCESS

The process by which the alignment between technology and strategy occurs was observed to be quite amorphous. Most literature which prescribes such a match takes a product or strategy centered view whereby appropriate technological choices are made to support a given strategy, customer, product life cycle, etc. (Hayes & Wheelwright, 1979; Hill & Duke-Woolley, 1983; Porter, 1983; Skinner, 1974). The entire underpinning of a focused factory is for the manufacturing capability to efficiently match competitive strategies. This pattern did indeed appear to be highly prevalent. M3, for example, found they had to adopt new process technology after changing strategies from short-run prototype and tooling jobs to large batch production work because the company was not competitive in this new market with their traditional machine tools. One CNC lathe was purchased as an experiment. Its success lead to a second, and then to machining centers, and so on, until 90 percent of production was on CNC equipment. The shift in strategies required a total shift in process technology. However, there were variations from the 'technology follows strategy' pattern as well as direct counter examples.

Many firms adopted CC technologies to support certain strategies but subsequently made opportunistic strategy deviations as they discovered that the technology came with unanticipated capabilities. Exploiting these newly found capabilities required adjustments in existing competitive strategies. T1, for example, discovered that new CNC equipment provided greater flexibility and capacity than anticipated. This allowed expansion into more specialized cutting tools:

FIGURE 3

PROPRIETARY PRODUCTS

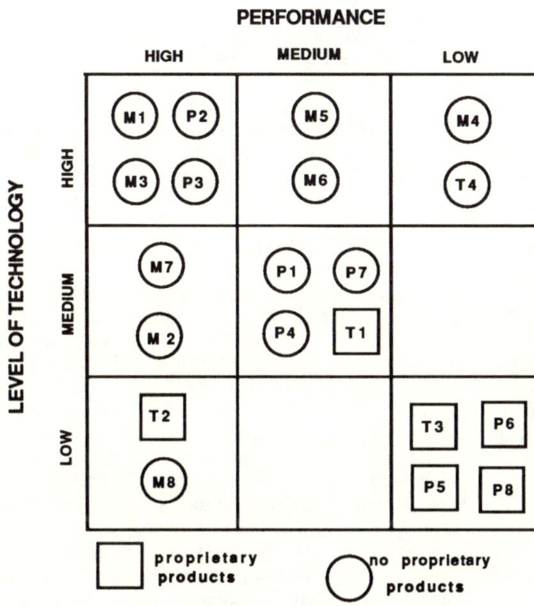

"We bought the machine thinking it's going to achieve certain things and are finding that we can do other things with it." (president T1)

Minor strategic adjustments to take advantage of unanticipated capabilities are consistent with Penrose's (1959) impetus for firm growth, where firms continually expand to take advantage of under-utilized resources. Exploiting these resources requires adjusting one's strategy.

The strategy alignment process deviated most from the traditional view when CC technology was purchased to support a specific strategy only to find there was a total mismatch. Three firms, T4, M2, and M8, found themselves in this situation. M8 corrected the problem by selling the new equipment, but T4 and M2 considered the adoption irreversible and radically altered strategies to fit the new technology. T4 transformed from a tool sharpener and regrinder to a primary tool manufacturer to take advantage of larger batch sizes. M2 hired a new sales person to search for work with longer production runs for the CNC equipment. This reaction is a form of process centered strategy (Buggie, 1981; Hughes, 1984) whereby customers and products are changed to remain compatible with a firm's large investment in process technology.

Overall, in light of the impact which manufacturing technology can have on strategy, it seems more useful to speak of a strategy-technology co-alignment process whereby strategy dictates technology and technology alters strategy. A major part of this alignment process takes place at the shop floor level where the new technology is being integrated into the manufacturing system.

DISCUSSION AND RESEARCH ISSUES

Linking manufacturing process technologies with generic competitive strategies proved to be problematic. Rivalry was primarily between firms with similar strategic profiles, possessing like capabilities, and competing with similar strengths (e.g. high quality, precision, fast delivery). Consequently, price became the primary competitive factor. The only exceptions to this rule were situations in which normal market forces were disrupted. Therefore, linking manufacturing strategies with competitive strategies appears useful only at the most general level because price competition dominates most situations.

Fortunately, the same forces limiting the role generic strategy plays opens another way to frame the linkage question. Groups of firms vying for different market segments can be expected to have different sets of capabilities and technologies. The importance of CC technology can be expected to vary greatly between groups. We observed that firms pursuing larger batch work, higher precision work, and doing contract jobbing were more reliant on CC technologies.

To classify rivals with similar competitive profiles, strategic management literature uses the construct of strategic groups (Harrigan 1985). Firms within a given strategic group have like characteristics and follow similar strategies. They are primary rivals for the same customers and markets at a more specific level than firms competing in an industry overall. The observation that an industry consists of "groups" of firms with like behavior was introduced by Hunt (1972) and Newman (1973), and rationalized when Caves and Porter (1977) identified mobility barriers within industries which keep groups separate.

Identifying groups for research purposes, however, has proven problematic. They are typically identified by either a priori industry knowledge or by examining data to identify key strategic variables which most clearly accentuate competitive similarities and differences (Porter, 1980; Harrigan, 1985). This involves a great deal of subjective judgement and provides no uniform means for cross industry comparisons. However, Cool & Schendel (1987) recently proposed that as a minimum standard, groups be classified in terms of business scope or domain and resource commitments, which are argued to be the two primary dimensions in pursuit of competitive advantage (e.g. Porter, 1985; Hofer & Schendel, 1978). Batch size of products, the level of precision required by customers, and the choice between proprietary or contracting work, all fit closely with traditional components of a firm's domain (Thompson, 1967; Abell, 1980). In addition, resource commitment is a primary element in technology choice decisions.

Strategic grouping offers great promise as a construct which can help link manufacturing technology more closely with competitive strategy. Indeed, many of Skinner's (1974) elements of a focused factory require specific domain and resource commitment decisions. Using strategic groups to further our understanding of technology - strategy linkages requires fuller development of the construct. Work is needed on identifying critical grouping dimensions. Such dimensions should be firmly grounded in theory and empirically validated in an effort to move beyond the current state of idiosyncratic industry specific groupings based largely on judgement and/or statistical techniques.

This study also observed that the traditional manufacturing strategy trade-off between cost efficiency and flexibility still holds in some cases, even with the advent of "flexible" technologies. Computer controlled technology, for all its acclaimed flexibility, came down on the side of cost/efficiency. Firms with larger - not smaller - batch sizes performed better with computer controlled technologies. This pattern was, however, observed to be shifting as the technology develops, particularly with more robust and user friendly software, an observation shared by others (e.g. Jelinek & Goldhar, 1983). This shifting pattern is first observed with firms that can gain most from the unique advantages of CC technologies.

The process by which process technology aligns with competitive strategy to form a "fit" was observed to follow several patterns. While most firms selected a technology to fit a strategy, there were often surprises once adopted. These ranged from new and unexpected capabilities to complete technology-strategy mismatches. The adjustment process that followed often involved changes in the firm's strategy to either exploit new capabilities or simply to survive. This raises important issues concerning factory focus. The adjustment to new capabilities can tighten technology-strategy fit or it can represent focus regression that Skinner (1974) and Hill & Duke-Woolley (1983) warn against. Maintaining focus while developing and pursuing emerging capabilities presents research questions on balancing focus and technology exploitation in a technologically dynamic environment. Just what are the critical strategies and dimensions of focus in today's "flexible" factory?

Due to the exploratory purpose of this study and its narrow, relatively static focus, care must be taken when interpreting its results. Many of the ideas and issues discussed deal with shifting situations driven by fast paced technological change. It is sometimes difficult to meaningfully examine these dynamic issues with traditional research methods. Too often we as researchers are late to begin work on important issues. By the time we reach the point of comfort about our intellectual rigor in research design, we may well be documenting history rather than discovering solutions to current problems. We must critically examine our research traditions in light of a changing world and the new needs of our constituencies. This may require modifying our own research culture and traditions to match the pace of an increasingly dynamic manufacturing environment. Techniques such as action research (Susman, 1983), in which scholars work real-time with managers as partners confronting manufacturing strategy challenges, offers attractive opportunities. This keeps researchers on the forefront of the field, ensuring we are working on the right questions at the right time while providing managers with assistance when it is most needed.

REFERENCES

Abell, D. Defining the Business. Englewood Cliffs, NJ: Prentice-Hall, 1980

Abernathy, W.J. "Production Process Structure and Technological Change." Decision Science, 7 (October 1976), 607-619.

Abernathy, W.J., and P.L. Townsend. "Technology, Productivity, and Process Change." Technological Forecasting and Social Change, 7 (1975), 379-396.

Abernathy, W.J., and J.M. Utterback. "Patterns of Industrial Innovation." Technology Review, 80 (June/July 1978), 2-9.

Blois, K. J. "Matching New Manufacturing Technologies to Industrial Markets and Strategies." Industrial Marketing Management, 14 (February 1985), 43-47.

Buffa, E.S. "Making American Manufacturing Competitive." California Management Review, 26 (Spring 1984), 29-46.

Buggie, F. D. "Reindustrialization on a Micro Scale." Business Horizons, (July-August 1981), 30-33.

Caves, R.E., and M.E. Porter. "From Entry Barriers to Mobility Barriers: Conjectural Decisions and Contrived Deterrence to New Competition." Quarterly Journal of Economics, 91 (May 1977), 241-261.

Chakravarthy, B. S. "Measuring Strategic Performance." Strategic Management Journal, 7 (1986), 437-458.

Congden, S.W., and D.M. Schroeder. "The Adoption of New Process Technologies: To Whose Best Advantage?" Working paper, University of Massachusetts, 1989.

Cool, K.O., and D. Schendel. "Strategic Group Formation and Performance: The Case of the U.S. Pharmaceutical Industry, 1963-1982." Management Science, 33;9 (September 1987), 1102-1124.

De Meyer, A., J. Nakane, J.G. Miller, and K. Ferdows. "Flexibility: The Next Competitive Battle - The Manufacturing Futures Survey." Strategic Management Journal, 10 (March-April 1989), 135-144.

Dertouzos, M.L., R.K. Lester, and R. M. Solow. Made in America: Rcgaining the Productive Edge. Cambridge, MA: MIT Press, 1989.

Dess, G.G. and R.B. Robinson. "Measuring Organizational Performance in the Absence of Objective Measures: The Case of the Privately-held Firm and Conglomerate Business Unit." Strategic Management Journal, 5 (1984), 265-273.

Fisher, F. and J. McGowan. "On the Misuse of Accounting Rates of Return to Infer Monopoly Profits." American Economic Review, (March 1983), 82-97.

Glaser, B. and A. Strauss. The Discovery of Grounded Theory. Chicago: Aldine, 1967.

Harrigan, K.R. "An Application of Clustering for Strategic Group Analysis." Strategic Management Journal, 6 (1985), 55-73.

Hayes, R.H., and R.W. Schmenner. "How Should You Organize Manufacturing." Harvard Business Review, 56 (Jan-Feb, 1978), 105-118.

Hayes, R.H., and S.C. Wheelwright. "Link Manufacturing Process and Product Life Cycles." Harvard Business Review, 57 (Jan-Feb 1979), 133-140.

Hayes, R.H., and S.C. Wheelwright. Restoring Our Competitive Edge: Competing Through Manufacturing. New York: Wiley & Sons, 1984.

Hill, T.J., and R.M.G. Duke-Woolley. "Progression or Regression in Facilities Focus." Strategic Management Journal, 4 (April-June 1983), 109-121.

Hofer, C.W., and D. Schendel. Strategy Formulation: Analytical Concepts. St. Paul, Minnesota: West Publishing Co., 1978.

Hughes, E.R. "Responding to Changes in Process Technology: Strategies for the Small Business."

Journal of Small Business Management, 22 (January 1984), 8-15.
Hunt, M. "Competition in the Major Home Appliance Industry, 1960-1970." Ph.D dissertation, Harvard University, 1972.
Jaikumar, R. "Postindustrial Manufacturing." Harvard Business Review, 64 (Nov-Dec 1986), 69-76.
Jelinek, M., and J. Goldhar. "The Interface Between Strategy and Manufacturing Technology." Columbia Journal of World Business, 18 (Spring 1983), 26-36.
Kotha, S., and D. Orne. "Generic Manufacturing Strategies: A Conceptual Synthesis." Strategic Management Journal, 10 (May-June 1989), 211-131.
Meredith, J. "The Strategic Advantages of New Manufacturing Technologies for Small Firms." Strategic Management Journal, 8 (May-June 1987), 249-258.
Newman, H. "Strategic Groups and the Structure-Performance Relationship: A Study with Respect to the Chemical Process Industries." Unpublished Ph.D dissertation, Harvard University, 1973
Penrose, E.T. The Growth of the Firm. Oxford: Basil Blackwell, 1959.
Piore, M.J., and C.F. Sabel. The Second Industrial Divide. New York: Basic Books, 1984.
Porter, M.E. Competitive Strategy. New York: Free Press, 1980.
Porter, M.E. "The Technological Dimension of Competitive Strategy." In Research on Technological Innovation, Management and Policy, Vol. 1, ed. Richard S. Rosenbloom. Greenwich: JAI Press, 1983.
Porter, M.E. Competitive Advantage. New York: Free Press, 1985.
Salamon, G. "Accounting Rates of Return." American Economic Review, (June 1985), 495-504.
Schroeder, D.M. "A Dynamic Perspective on the Impact of Process Innovation Upon Competitive Strategies." Strategic Management Journal, 10 (Nov-Dec 1989).
Schumpeter, J. A. The Theory of Economic Development. Cambridge, MA: Harvard University Press, 1934.
Skinner, W. "Manufacturing -- Missing Link in Corporate Strategy." Harvard Business Review, 47 (May-June, 1969), 136-145.
Skinner, W. "The Focused Factory." Harvard Business Review, 52 (May-June 1974), 113-121.
Skinner, W. Manufacturing: The Formidable Competitive Weapon. New York: John Wiley and Sons, 1985.
Stobaugh, R., and P. Telesio. "Match Manufacturing Policies and Product Strategy." Harvard Business Review, 61 (March-April, 1983), 113-120.
Susman, Gerald I. "Action Research: A Sociotechnical Systems Perspective." In Beyond Method: Strategies for Social Research, ed. Gareth Morgan. Beverly Hills, CA: Sage, 1983.
Swamidass, P.M. "Planning for Manufacturing Technology."
Long Range Planning, 20 (October 1987), 125-133.
Thompson, H., and M. Paris. "The Changing Face of Manufacturing Technology." Journal of Business Strategy, 3 (Summer, 1982), 45-52.
Thompson, J.D. Organizations in Action. New York: McGraw-Hill, 1967.
Utterback, J.M. "The Dynamics of Product and Process Innovation in Industry." In Technological Innovation for a Dynamic Economy, ed. Christopher T. Hill and James M. Utterback. New York: Pergamon, 1979.
Voss, C.A. "Managing New Manufacturing Technologies." Operations Management Association, Monograph No. 1, Michigan State Univ., Sept. 1986, 1-68.
Wheelwright, S.C. "Reflecting Corporate Strategy in Manufacturing Decisions." Business Horizons, 21 (Feb 1978), 57-66.
Wheelwright, S.C. "Manufacturing Strategy: Defining the Missing Link." Strategic Management Journal, 5 (1984), 77-91.
Williamson, O.E. Markets and Hierarchies. New York: Free Press, 1975.

18

DIMENSIONING THE SERVICE FACTORY

Richard B. Chase

Center for Operations Management, Education, and Research
School of Business Administration
University of Southern California
Los Angeles, California 90089-1421
(213) 743-8338

ABSTRACT

The Service Factory has been proposed as a new approach to manufacturing strategy. This paper reviews the information, problem-solving, sales, and support dimensions which underlie a service factory. It also provides a strategic planning methodology by which it can be applied, and raises five research questions which might be addressed through empirical study.

INTRODUCTION

Two recent articles (Chase and Erikson [1988], Chase and Garvin [1989]) have proposed that manufacturing organizations can become more competitive by employing a broader range of services provided by factory personnel and facilities. The general concept has been termed the "service factory," in which service is identified as the "fifth competitive priority" in manufacturing strategy. It is the purpose of this paper to review the service factory concept, provide a first-cut strategic planning methodology to apply it, and to raise some research questions relating to its further development as a distinctive manufacturing strategy.

Review of the Service Factory Concept

The concept of a well-performing factory has evolved over time. Traditionally, the emphasis has been on material transformations and the production of tangible products. Raw materials and components were the inputs; parts, subassemblies, and finished goods were the outputs. A well-performing factory was one that accomplished these transformations at low cost, and with high quality. In recent years, attention has shifted to flexibility: the range of transformations or products that the factory can accommodate, and the speed with which changes can be made. A well performing factory is now judged not only by the efficiency of its operations, but also by their scope and ease of change.

Factory service is underemphasized in both of these models. In traditional factories, service has usually been defined narrowly: as speed and dependability of delivery, or as accuracy and completeness of order fulfillment. Typically, narrow and partial measures have been used to evaluate success of these criteria. In flexible factories, the concept of service has broadened to include both breadth of product offerings and the ability to customize to meet specific needs. However, even flexible factories fail to recognize the full range of services that factories can provide. In addition to delivery, order fulfillment, and flexibility, they include:

Information: the ability to furnish critical data on product performance, process parameters, and cost to internal groups, such as R & D, and to external customers, who then use the data to improve their own operations or products (example: Hewlett-Packard's Fort Collins quality department provides product quality data sheets to field sales and service personnel, as well as videotapes documenting actual product testing and field quality performance);

Problem-solving: the ability to assist internal and external groups in problem-solving, especially in such areas as new product development, design for manufacturability, cost reduction, and quality improvement (example: Raritan Corporation, a metal rod fabricator, sends factory workers out with salespeople to troubleshoot quality problems; they then return to the factory and work with shop floor personnel on remedial efforts);

Sales: the ability to enhance sales and marketing efforts by showcasing, in real-time, the technology, equipment, or production systems the company is trying to sell (example: Digital Equipment Corporation sells its CIM system, in part, by showcasing it in action on its own factory floor);

Support: the ability to provide customers with quick replacement of defective parts (example: Caterpillar's promise to have repair parts available anywhere in the world within 48 hours), or quick replenishment of stocks to avoid downtime or stockouts (example: Beneton, a retail chain, is linked to its Italian textile mills via a sophisticated computer system that allows the factories to begin producing items that are selling well as soon as weekly sales figures have been collected).

These activities, which can be referred to as IPSS services, lead to a new manufacturing concept: the service factory. A service factory has two defining characteristics:

1. *The factory provides a range of services as well as goods, and it recognizes the combination.* Most factories today are struggling to become more customer-oriented. Few, however, appear to have recognized that their customers--both internal and external--are often as interested in the services that accompany products as they are in the products themselves. Exhibit 1 lists, by customer group, the services desired by the factory's internal and external customers. The Exhibit prompts several observations:

a. The range of desired services is far broader than most manufacturing executives recognize. Delivery and availability remain important, but information, problem-solving, sales, and support (IPSS) dominate the list.

b. The factory's customers are both upstream and downstream in the chain of value-creating activities. This is in marked contrast to the perspective of modern quality control, which argues that "the next (downstream) step in the process is the customer."

2. *The factory is managed so that these services are a central focus; they are explicitly measured, monitored, and marketed.* The manufacturing strategy literature suggests that managers should choose a dominant goal from among four priorities: cost, quality, delivery, and flexibility. The service factory, however, presents a fifth possible focus: service, defined as some combination of information, problem-solving, sales, and support. The successful provision of these services requires a shift in attitudes. Traditionally, manufacturing managers have sought to seal off their core production technology from outside disturbances, in order to maximize efficiency. A closed systems logic has ruled, with physical (i.e., inventory) and organizational buffers erected between the factory and its customers. Service factories, however, are based on a different logic: an open systems philosophy, with direct and accessible linkages between the factory and its internal and external customers. This philosophy requires that factories be designed to maximize the effectiveness of their operations in terms of the benefits provided to customers (See Exhibit 2).

Service factories also feature service goals more prominently in their measurement systems, mission statements, and day-to-day operations. Services are no longer simply an add-on, secondary to the product being offered, but are viewed as an integral part of operations design, planning, and execution.

Despite parallels to traditional and flexible factories, service factories are distinctive in several ways. Service factories are more closely coupled with their internal and external customers than traditional factories, and view service far more broadly (See Exhibit 3). At traditional factories, service is little more than meeting due dates. Information exchange is limited. The factory is buffered from the outside world by other departments (especially sales, marketing, and engineering), and customers are, all too frequently, visualized as numbers on a production schedule.

Flexible factories are more closely coupled to customers than traditional factories, but the focus remains narrow: rapid design changes and volume adjustments. At times such responsiveness is achieved through advanced technology (e.g., FMS, robotics); more often the approach is continual scrambling and schedule readjustments. Either way, the result is a limited conception of service, and one that fails to capitalize fully on the factory's full range of service offerings. By contrast, service factories establish direct and continual linkages between the production organization and internal and external customers. The factory and its systems are explicitly designed to be accessible, easy to work with, and responsive to customers' service needs. A broad range of IPSS services may be provided. In practice, however, most service factories tend to operate in one of four modes, each with its own mission, objectives, and skills:

1. the laboratory

2. the consultant
3. the showroom
4. the dispatcher

When a service factory is predominantly a laboratory, the focus is on information, problem-solving, and data collection. Pilot plants are the classic example of service factories operating in this mode. Their primary objectives are the testing of new products and processes to ensure manufacturability and the feedback of critical information to R & D. To meet these goals, processes are heavily instrumented, equipment is capable of fine adjustments, operators are technically skilled, and documentation is careful and precise. Occasionally entire factories will operate in a laboratory mode. New production methods will be tested in real-time, on operating equipment. To ensure success, innovative systems or highly trained operators are required.

When a service factory is predominantly a consultant, the focus is on problem-solving. Problem-solving by factory personnel can take various forms. For internal departments like R & D, it can mean assistance with new designs or advice on manufacturability; for external customers, it can mean help in understanding the performance limits of a part or subsystem or help in improving product quality. (Note that these activities must be performed by *factory* personnel, rather than by applications engineers or other groups, if the organization is to qualify as a service factory.) To meet these goals, factory personnel must have skills in problem-solving. They must be knowledgeable about their products and production processes, trained in basic statistical techniques, and comfortable dealing with customers. ACS, an Eli Lilly Subsidiary, has doctors come to their plant not only to demonstrate how their operating equipment is manufactured, but to enable the MD's to make their own equipment. For example, angioplastacy parts must often be calibrated in special ways to match the particular operating procedures use by the individual surgeon. ACS, through the use of video tapes and specially trained technicians, trains visiting surgeons in product assembly and testing. In the same spirit, Batesville Casket brings customers (mortuary owners) to their plant to see how the product is made and to permit on-the-spot customization of caskets.

When a service factory is predominantly a showroom, the focus is on marketing and sales. Factories can be a showroom in either of two ways:

a. They can provide a working demonstration of the systems, processes, and equipment that the company is selling;

b. They can dramatize the company's manufacturing superiority, and, by implication, its superior quality and reliability.

We call the first of these approaches "modeling your own clothes." It is used increasingly by companies that specialize in some aspect of computer-integrated manufacturing (CIM). For such companies, the real challenge is convincing potential users of the viability and ease of operation of their systems. Frequently, this can only be accomplished by on-site demonstrations and explanations. A variant of "modeling your own clothes" is the use of factory tours to reinforce product positioning. At Toyota, posted SPC shop paper visually documents quality; at Ben & Jerry's factory tours highlight those stages of the production process in which "chunks" are added to the ice cream, which highlights the company's distinctive marketing strategy.

A factory can also be a showroom for superior manufacturing skills. In that case, the service provided to marketing is not a direct product demonstration, but the inference that "if our factory is this advanced, you can bet that our product is too." Such indirect evidence is particularly important in selling capital goods, whose reliability and durability may take years to demonstrate.

Whichever of these two approaches is used, several requirements must be met if the factory is to be a successful showroom. Marketing and manufacturing personnel must work together closely to ensure that factory tours meet customers' needs; manufacturing managers and shop floor employees must be trained in communication and presentation skills; and the factory's design and layout must highlight those aspects of the product or process having the greatest sales potential.

When a service factory is predominantly a dispatcher, the focus is on logistics and field support. In this mode, the primary service goal is responsiveness. Quick replenishment of stocks and quick replacement of defective parts are especially critical. The factory tries to minimize the amount of time that intermediate customers (i.e., retailers and wholesalers) are out-of-stock, and also to maximize the product up-time of final customers. Traditional factories, of course, have similar goals. However, they typically meet these needs by carrying large inventories or by expediting a large proportion of their orders (JIT shops not withstanding), rather than by creating tighter linkages with customers and more responsive manufacturing organizations. Tighter customer linkages can be achieved in several ways. One is through the use of interorganizational systems (IOS) where customers are supplied with computer terminals that are linked directly to the factory's order entry and production control systems. Tighter customer linkages can also be achieved more conventionally, without sophisticated information systems. Typically, interfunctional teams are required, consisting of representatives from sales, logistics, and manufacturing. These teams meet regularly and share common goals. In any case, the key to becoming a responsive dispatcher is customer information. Factory managers need to understand order patterns, the assumptions behind forecasts, and special sales needs in the same depth as sales and marketing personnel. If interorganizational systems are used, customers' systems must be closely meshed with existing order entry and production control systems so that compatibility is ensured. Refined measures must be developed to track responsiveness, and customers must be polled to ensure that uniform standards (e.g., "deliver all parts within three days") are not being employed to meet needs that, in fact, differ. Sales, marketing, and manufacturing personnel must, of course, communicate on a regular basis.

Strategic Planning for the Service Factory

Before a service factory can be developed, new attitudes must be firmly in place. There are two primary requirements:

1. A clear understanding that the factory provides not only physical products, but a bundle of service benefits to its internal and external customers.

2. A recognition of the wealth of specific IPSS services that the factory could concievably offer its customers.

Once the proper attitudes are in place, the following four steps are suggested:

1. *Survey internal and external customers to assess their service needs.* Internally, this might involve plant managers sitting down with R & D groups to get them to specify useful feedback when using the factory as a laboratory for new processes or products. It might also involve meetings between plant managers, marketing managers, and sales and service managers to identify the kinds of services that are already being offered to these functions by the factory, as well as discussions with factory supervisors to obtain a listing of the kinds of IPSS services their work units currently provide. Externally, manufacturing managers might ask marketing to set up joint meetings with sales personnel, marketing personnel, and customers to identify what service means to end users. In addition, the customer service department might administer simple service questionnaires to a broad sample of outside customers. In both cases the customer would be asked to identify what information, problem-solving, and support services were important to them.

2. *Develop a service mission for the plant.* This entails a comparison of the factory's performance on the desired IPSS services, identified in step 1, with the performance of competitors. Here, management must decide whether it wishes to be world-class, to match the service performance of other manufacturers, or to lag in service. It must also decide on the aspects of service that it wishes to emphasize. If service is to be emphasized, managers should begin by issuing a general call to arms: for example, a statement such as "Our factory's mission is to become recognized equally for service and product quality." Managers must then specify the details of the factory mission in terms of service roles and goals. Although combinations of factory roles are possible, it is suggested, at the outset, that the focus be on a single mission. Thus, if the factory is to be a laboratory, the service mission would be to gather process data quickly and effectively to support R & D; if it is to be a consultant, the mission would be to help customers, both internal and external, solve quality and applications problems; if it is to be a showroom, the mission would be to contribute to sales and marketing by demonstrating the capabilities of products and processes; and if it is to be a dispatcher, the mission would be to provide timely and accurate information in a form desired by customers.

3. *Formalize the service factory strategy.* It is recommended that a one page statement be developed that can be folded into the plant charter. A typical plant charter describes

what is expected of a plant and the capabilities that it should have. The subjects normally covered include products and volumes to be produced, what the plant must do well, process capabilities and capacities, and directions for future development. Each of these would have to be modified if IPSS services are to become more central to plant performance. Implicit in the development of a service strategy is the need to define the scope and resultant costs of the services to be provided. Priorities have to be established among the needs of internal and external customers, and cost-benefit analyses conducted. Companies that decide to pursue an aggressive service strategy may well incur significant training, MIS, cost accounting, and facility modification costs, depending upon which of the IPSS dimensions they choose to emphasize.

4. *Develop detailed implementation plans.* It is suggested that management start with a single, high impact service--i.e., a service with considerable visibility to either the factory's internal or external customers, and likely to offer a substantial payoff. For example, if a company makes custom components for OEMs, the service might be a new job tracking system that gives customers quick feedback on the status of work orders. If a company relies heavily on new products, the service might be an innovative instrumentation program for the plant to provide R & D with test results that match the forms required for process certification. And if a company sells integrated hardware and software systems, the service might be an orchestrated plant tour to help win big orders.

It is essential to include employees early in the implementation planning process, because they are the ones ultimately responsible for service delivery. In fact, if used properly, the service factory can be a powerful HRM tool. An appreciation of one's skills and knowledge has long been recognized as a primary motivator of employees. By making the factory a showroom of skills as well as a showroom of equipment, the service factory allows workers to demonstrate their abilities to the firm's most important audience--its customers. We therefore see the service factory as a natural complement to continuous improvement and other quality and productivity programs. As such, it can be easily addressed by placing it on the agendas of existing quality and productivity improvement teams.

After the service delivery process has been designed, it is important to set up a meeting of department managers from sales, R & D, personnel, production, and MIS, in order to link the factory's service objectives with performance measures. Since the service innovations proposed span departmental boundaries, cross-functional support is a basic requirement for successful implementation--involved departments must operate as "good customers." For example, marketing must let the factory know what it is promising customers in the way of factory-based services. Departments which are service providers, such as MIS and personnel, should help provide the information systems and training to enable the factory to excel at its service mission. And, of course, senior management must ensure that the desired service objectives and their corresponding measures are given proper weighting in annual performance reviews.

Specific service performance goals must also be established. It is not enough to say that "service comes first," or that "superior service is essential for long term success." Few managers or employees will act on such vague pronouncements. It is far more meaningful to require that response time for service requests not exceed two hours, or that X% of all customers will rate their service encounters with the factory as "excellent". (See Exhibit 3 for additional service measures).

Finally, monthly meetings should be held to review the service aspects of factory performance. Customers should be included in these meetings, and continually asked two questions: (1) What type of help would you like from the factory? and (2) How can our factory operations be made more user friendly?

RESEARCH QUESTIONS

The broad question is: How do service factory objectives affect technology decisions, personnel decisions, and structural decisions of the manufacturing organization?

Narrower questions which fall out of this are:

1. What is the nature and extent of tradeoffs that exist between IPSS services and traditional manufacturing priorities? What do you have to give up to compete on service?

2. Is there a relationship between service factory services and corporate performance measures such as market share and profitability?

3. Does a company have to be world class in its manufacturing operations before it can successfully compete through any or all of the IPSS services?

4. What is the "best practice" in performing IPSS services? Offering such services is not necessarily the same as doing them well. This point is indicated graphically in Exhibit 4, where service is seen as a two dimensional construct consisting of scope or range of IPSS services attempted, and quality of the IPSS services provided. A factory that provides neither a wide range of services nor particularly high quality in the ones it does deliver is labeled as a "base factory;" a factory that does a good job on a narrow range of services is a "flexible factory;" a factory that tries to perform a wide range of services, but does so ineffectively is labeled a "scramble factory," and a factory that does a good job across a wide range of services fits the conception of a complete "service factory." Clearly, the research question here is specifying explicit criteria for achieving high quality in each service offered.

5. Relative to Exhibit 4, what is the appropriate path to follow to introduce service factory capabilities into a manufacturing organization? Starting with a base factory, it is obvious that one would not plan to become a "scramble factory" on the way to the valhalla of the service factory. But what about becoming a flexible factory first, or concurrently, in striving toward the service factory? (Wheelwright and Hayes (1985) talk about making "the great leap" to stage 4 manufacturing. Is the

same analysis they present appropriate when considering moving directly to a service factory?)

CONCLUSION

Is the service factory really a new concept, or just a new wine in an old bottle? Perhaps it really doesn't matter. That is, it would seem that the true test of value is whether looking at manufacturing strategy with a clear conceptualization of service uncovers opportunities for competitive advantage. The fact that many of our outstanding manufacturers have some elements of the service factory evidenced in their operations is proof that a service based strategy is viable. It will be interesting to see how service based competition arises in the battle for Europe in 1992. It will also be interesting to see how the Japanese compete on service. One thing is for sure--they will.

REFERENCES

Chase, R. B. and Erikson, W. J., "The Service Factory," *Academy of Management Executive,* Vol. II, No. 3, August 1988, pp. 191-196.

Chase, R. B. and Garvin, D. A., "The Service Factory," *Harvard Business Review,* Vol. 67, No. 4, July-August 1989, pp. 61-69.

Wheelwright, S. C. and Hayes, R., "Competing Through Manufacturing," *Harvard Business Review,* Vol. 67, No. 4, July-August 1989, pp. 61-69.

Exhibit 1

The Range of Factory Services

Services Desired by Internal Customers			Services Desired by External Customers	
Research and Development	Logistics and Distribution	Sales and Marketing	Wholesalers and Retailers	End Users
accurate and feedback on the manufacturability of new designs	timely and complete order fulfillment	real-time demonstrations of the technology being sold	timely and complete order fulfillment	timely and complete order fulfillment
rapid construction of prototypes	packaging for ease of shipment and minimal handling damage	responsiveness to promotions and other special needs	accurate and accessible information on order status	accurate and accessible information on order status
smooth introduction of engineering change orders	advance notice of delivery problems	a wide range of product offerings and the ability to customize	direct computer linkages to facilitate purchasing	direct computer linkages to facilitate purchasing
		detailed cost and production data for competitive bidding and cross-selling		assistance with installation maintenance, and problem-solving
				the ability to change the content and timing of orders on short notice

Exhibit 2

CLOSED SYSTEM MODEL

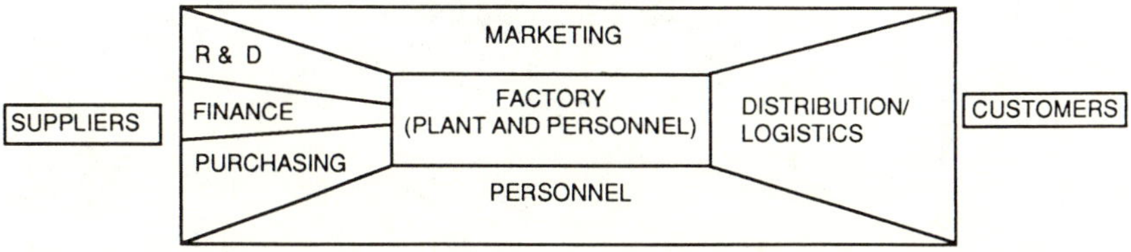

MANUFACTURING IS SEEN AS AN INTERNAL FUNCTION WHICH IS BUFFERED FROM THE CUSTOMER IN ORDER TO MAXIMIZE EFFICIENCY

OPEN SYSTEM MODEL

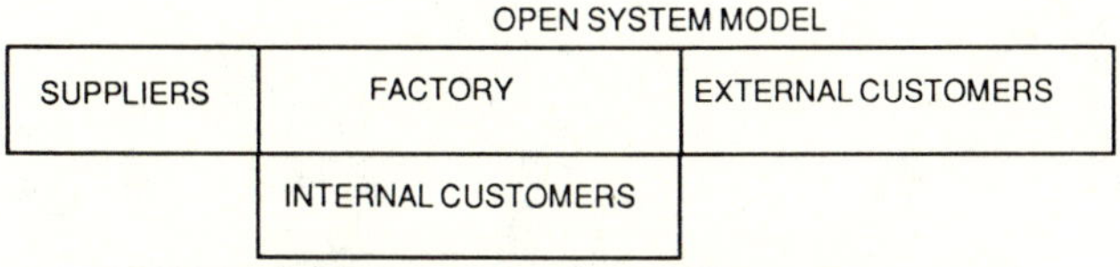

MANUFACTURING IS CLOSELY LINKED TO INTERNAL AND EXTERNAL CUSTOMERS TO MAXIMIZE EFFECTIVENESS

Exhibit 3

Contrasting Models of the Factory

	Traditional Factory	Flexible Factory	Service Factory
Mission and Role	To provide products in a timely fashion.	To provide a wide range of products in a timely fashion.	To provide products and there associated services in a timely fashion.
Service Offerings	Delivery, availability.	Delivery, availability customization.	Delivery, availability, customization, information, problem-solving, sales, support.
Services Measures	% on-time delivery, % orders shipped on promised dates, % orders shipped complete.	% on-time delivery, % orders shipped on promised dates, % orders shipped complete, range of product offerings speed of response to requested product changes.	% on-time delivery, % orders shipped on promised dates, % orders shipped complete, range of product offerings, speed of response to requested product changes, speed of response to engineering requests, speed of response to customer emergencies, speed and completeness of response to data inquiries, customer satisfaction with problem solving and sales assistance.
Required Service Skills	Production control scheduling.	Production control, scheduling, responsiveness to product and volume changes.	Production control, scheduling, responsiveness to product and volume changes, order tracking, data collection, technical problem-solving, knowledge of customer needs, ability to interact with customers.

EXHIBIT 4. Service Scope and Service Effectiveness

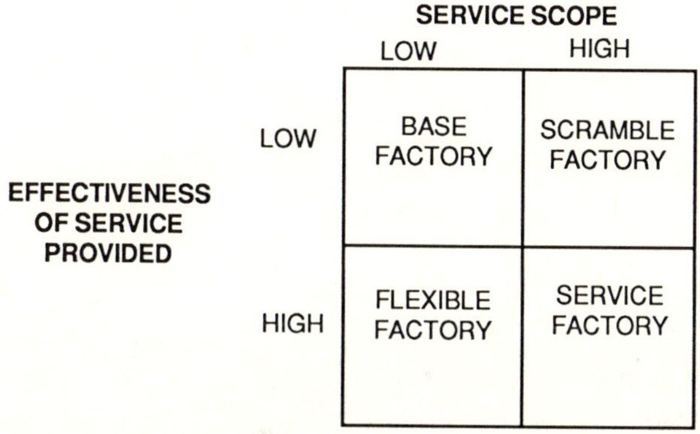

19

MANUFACTURING STRATEGY: AN OVERVIEW OF CURRENT PROCESS AND CONTENT MODELS

Peter T. Ward, G. Keong Leong and David L. Snyder

College of Business
The Ohio State University
Columbus, OH 43210
(614) 292-5294

ABSTRACT

In this paper, a content and process model of manufacturing strategy is developed based on our interpretation of the literature. While we argue that both models are inadequate, we conclude that the content model is better established than the process model. We also argue that progress toward a paradigm of manufacturing strategy has been slow. Three main reasons for this lack of progress are offered: a paucity of explicit attempts to build theory, a shortage of empirical research, and too little integration of manufacturing strategy research with research in related disciplines. Suggestions are made to address each of these shortcomings.

INTRODUCTION

Although still a relatively new area of study and practice, most writers trace the beginning of manufacturing strategy as a field to Wickham Skinner's (1969) early conceptual work. During the 1970's the field consisted of the writing of relatively few authors and there were few outlets for publication, notably the ***Harvard Business Review***. The conceptual work published in the 1970's encouraged many new authors to join the fray and the numbers of articles and outlets for manufacturing strategy writers have increased during the 1980's. Despite these increases, few attempts have been made to review, organize and criticize the emerging base of manufacturing strategy literature. Adam and Swamidass (1989), Anderson, Schroeder and Cleveland (1989) and St. John (1986) are notable exceptions.

In this paper, we suggest two complementary models for viewing manufacturing strategy and research. We also critically evaluate these models with respect to the published research. Finally, we identify gaps in the research and offer our view of a research agenda for the early 1990's that addresses these gaps.

DICHOTOMY OF MANUFACTURING STRATEGY RESEARCH

Business strategy researchers have long noted the distinction between research on the **process** of strategy and research on the **content** of strategy (Huff and Reger, 1987; Fahey and Christiansen, 1986). Until very recently, this distinction has not been made explicitly in manufacturing strategy research. Adam and Swamidass (1989) assess several major contributions to the manufacturing literature on the basis of process and content variables used in the studies. This process-content distinction provides useful organizing categories for manufacturing strategy research.

The process of determining manufacturing strategy and its content have both been addressed in the manufacturing strategy literature, although far more emphasis has been given to content of strategy than to the process by which it has been made. As will be discussed, broad agreement exists among the various authors with respect to both the process by which manufacturing strategy should be made and the appropriate content of manufacturing strategy.

Process Model

Skinner (1969) introduced the concept of manufacturing strategy's role in the corporate framework. Along with the efforts of Hayes and Wheelwright (1984), Anderson et al. (1989), Fine and Hax (1985), Hill (1989), and Roth (1989), Skinner's work suggests a hierarchical model in which corporate strategy drives business strategy which, in turn drives the strategies of manufacturing and other functional areas within the business unit (e.g., marketing, research and development). Although the "consensus" model is hierarchical, Hayes (1985) cautions that functional capabilities should, in some cases, drive corporate strategy.

The model presented in Figure 1 is, in our view, an amalgam of the work of the authors cited above. The model depicts the strategy process occuring within an environment that consists of markets (competitors, customers, and financial), and stakeholders (the public at large, stockholders, government agencies, and others). Corporate strategy and business unit strategies determine an appropriate pattern of functional strategies for each business unit, with feedback on functional capabilities (or distinctive competences) provided throughout the process.

The functional strategies themselves (see Figure 1) consist of strategic planning, implementation and capabilities which may, if the strategy is realized, result from planning and implementation. This model does not preclude the possibility that capabilities may also result from unplanned patterns of activities rather than a strategic plan. Miller and Hayslip (1989) assert that capability development and strategic planning are activities which should be jointly undertaken in order to achieve a competitive advantage.

The collective effort and capabilities of the manufacturing business's functional areas serve to produce a service-enhanced product. A service-enhanced product is simply the bundle of goods and services available to the customer upon purchase of the product. For example, a product with certain physical characteristics, user-training, convenient packaging, logistic support, etc., would meet this definition because the manufacturing, marketing, distribution and field service operations all contribute to the bundle being purchased by the customer. Thus, the service-enhanced product is the result of the efforts of all of the functional areas. Chase and Garvin (1989) use examples which demonstrate the multi-functional nature of producing service enhanced manufactured products and provide a strong argument for the importance of considering service and product together.

Next, internal performance measures reflect how well the service-enhanced product meets strategic objectives. These measures are depicted following the service-enhanced product to emphasize the importance of the "fit" of the capabilities brought to bear by each function. Dixon, Nanni, and Vollmann (1989) have promoted the importance and relative neglect of meaningful systems of performance measurement in introducing their Strategy-Action-Measurement (SAM) model. Johnson and Kaplan (1987) document the problems inherent in cost accounting metrics for measuring manufacturing performance.

Finally, the market itself provides a telling, external performance measure of the service-enhanced product and, hence, strategy. Hill (1985, 1989) puts particular emphasis on the importance of order-winning and the marketing function in his process model of manufacturing strategy. The need for internal and external performance measures is consistent with the suggestion of Garvin (1983).

It should be reiterated that the model presented in Figure 1 is our interpretation of the literature's dominant manufacturing strategy process model. This model, in our view, has several

Figure 1: A Process Model of Manufacturing Strategy

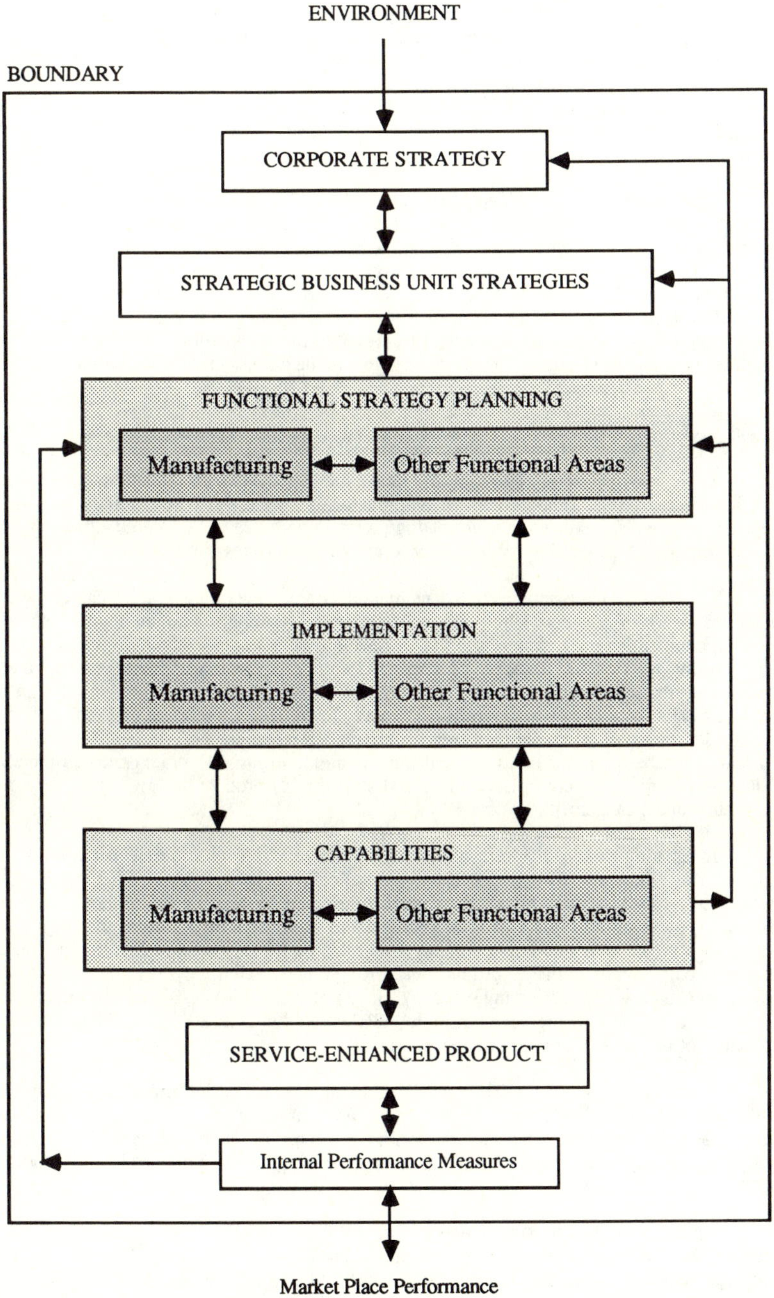

weaknesses. First, the model suggests a hierarchical neatness that has not been demonstrated to exist in practice. As noted, Hayes (1985) has argued that functional capabilities may determine strategy. This suggests a bottom-up (or middle-up and down) approach to strategy making that has not been adequately considered in the literature. Second, the adequacy of the manufacturing strategy process itself has been questioned. Miller and Hayslip (1989) argue that manufacturers should balance the use of both strategic planning and capability building to achieve competitive advantage. The implication of their argument is that the process model should be revised for parallel development of conventional strategic planning activities along with capability building activities such as "total quality control".

In short, research is needed to test the adequacy of this model to capture the essential nature of how manufacturing strategy is developed. Of course, others may disagree with this characterization of the published research on the manufacturing strategy process. However, we contend that a basic understanding of the manufacturing strategy process has not been reached, nor has the subject received adequate debate or field research. Therefore, we view the process model presented in Figure 1 as only a starting point for understanding how manufacturing strategy is made and how it fits with other strategic efforts within the corporation. A perspective of **what** constitutes manufacturing strategy is provided below in the discussion of a content model.

Content Model

There is essential agreement in the field's published work on the appropriate content of manufacturing strategy. The content of manufacturing strategy can be captured in two broad categories: 1) **decision areas** that are of long term importance in the manufacturing function, and 2) **competitive priorities** based on corporate and/or business unit goals.

The conceptual beginnings of a content model of manufacturing strategy can be credited to the seminal work of Skinner (1969). Skinner conceived a model for manufacturing strategy in which the competitive environment suggests a basic business strategy which, in turn, suggests the manufacturing mission or strategy. This mission can be encapsulated into choices made with respect to four competitive priorities: cost, quality, delivery and flexibility. The design of the manufacturing system can be made to fit the strategy by making appropriate tradeoffs or decisions in certain key areas. Further, Skinner suggested five areas where tradeoff decisions had to be made to assure a fit between business strategy and manufacturing: 1) plant and equipment; 2) production planning and control; 3) labor and staffing; 4) product design/engineering; and 5) organization and management.

In ensuing years, different authors have offered lists of strategic decision areas for manufacturing (Skinner, 1974; Buffa, 1984; Hayes and Wheelwright, 1984; Fine and Hax, 1985; Hayes et al., 1988). Table 1 compares the lists offered by four of the authors cited above. Essentially, these authors' lists of strategic decision areas converge to ten; the original eight adopted by Hayes and Wheelwright, a ninth (suggested by Buffa, Fine and Hax, and Hayes et al.) concerned with the fit between product and process and their lifecycles. A tenth category, performance measurement, is suggested by Hayes et al. (1988) as an important element in developing a manufacturing infrastructure. Research by Dixon et al. (1989) stresses the importance of decisions about performance measures.

A study by Ward, Miller and Vollmann (1988) based on Boston University's Manufacturing Futures data empirically derived areas of strategic concern expressed by managers. These areas were found to match the decision areas offered in Table 1 quite well. The study by Ward et al. provides some empirical validation for the conceptual work offered by the authors listed earlier.

An important addition to the lists of strategic decision categories suggested by the various authors is the structure - infrastructure bifurcation developed by Hayes and Wheelwright (1984). Structural decision categories address the "bricks and mortar" decisions of capital spending. Infrastructural decisions affect the people and systems that make manufacturing work. Hayes et al.

(1988) suggest that the distinction between structure and infrastructure is analogous to the distinction between computer hardware and software. Further, they suggest infrastructural decisions are equally important to structural decisions. Hill (1989) picks up on their theme and uses it to help organize his pioneering text.

A review of the literature on competitive priorities suggests a level of agreement similar to that found with strategic decision areas. A composite view (Skinner, 1969, 1978, 1985; Wheelwright, 1978, 1981, 1984; Buffa, 1984; Hayes and Wheelwright, 1984; Fine and Hax, 1985; Hayes, 1985; Van Dierdonck and Miller, 1980; Hayes et al., 1988) yields the following five dimensions of competitive priorities:

1) Cost;
2) Delivery Performance -- dependability and speed;
3) Quality;
4) Flexibility -- product mix and volume; and
5) Innovativeness.

Huete and Roth (1987), using data from the 1984 Boston University Manufacturing Roundtable Survey, derived factors which are representative of the first 4 competitive priorities listed earlier.

Table 1: Comparison of Different Authors' Lists of Strategic Decision Categories

	Skinner	H & W**	Buffa	Fine & Hax
STRUCTURAL	• Plant & equipment	• Capacity • Facilities • Technology • Vertical integration	• Capacity/location • Product/Process technology • Strategy w/ suppliers vertical integration	• Capacity • Facilities • Processes & technologies
INFRASTRUCTURAL	• Production Planning & Control • Organization & management • Labor & staffing • Product design/ engineering	• Production planning & control • Quality • Organization • Workforce • New product development * • Performance measurement systems *	• Strategic implications of operating decisions • Workforce & job design • Position of production system	• Product quality • Human resources • Scope of new products

Note: * Indicates addition by Hayes et al. (1988); ** Hayes and Wheelwright.

Early schools of thought on competitive priorities implied trade-offs between cost and quality or between dependability and flexibility (Skinner, 1969; Banks and Wheelwright, 1979; Miller, 1983). Japanese manufacturers have shown that a choice between cost and quality is unnecessary, instead the real issue is in identifying techniques to improve quality that will reduce overall cost (Wheelwright, 1981). Ferdows et al. (1986) note that the Japanese manufacturers, having attained success in both quality and delivery management, are pursuing a new competitive strategy by focusing on combining low cost manufacture with flexibility. This observation is confirmed by De Meyer, Nakane, Miller and Ferdows (1989) manufacturing futures survey where the Japanese manufacturers placed cost at the top of their list of competitive priorities, ahead of rapid design changes and quality. This does not mean that quality is of lesser importance to the Japanese manufacturers but rather that "competitive priorities do not have to be the reflection of existing strength" (De Meyer et al., 1989). Instead of viewing priorities as alternative stand-alone points of concern, the Japanese manufacturers sequentially consider quality, dependability, cost and flexibility as priorities that a firm addresses over time.

Summary

As discussed, there is general agreement on the content of manufacturing strategy. The content of manufacturing strategy has been subjected to much more conceptual development in the literature than process issues. In addition, empirical studies have most often focused primarily on content issues. For example, Swamidass and Newell (1987) and Ward et al. (1988) are all, in essence, content studies. As such, the content model of manufacturing strategy is better defined than the process model, but still requires much research. In particular, the interactions among the various content elements are not well understood.

Our purpose in examining the process and content of manufacturing strategy has been to: 1) articulate the distinction between process and content research, 2) evaluate the dominant process and content models, and 3) examine the need for additional research from both a process and content perspective. With the process-content distinction in mind, we now propose a research agenda for the manufacturing strategy field.

RESEARCH AGENDA

Our review of the literature in manufacturing strategy reveals an increasing level of interest in the field but no similar acceleration in progress toward a well organized paradigm. In our view, this lack of progress results from three causes. First, there has been a relative dearth of cohesive theory building efforts on the part of manufacturing strategy scholars. Many of the best efforts have been aimed primarily at practicing managers and, as a result, have not been rigorous in reviewing and synthesizing previous efforts. In fact, peer reviewed journals which enforce such rigor have not been a major outlet for manufacturing strategy literature. A second cause for the lack of progress has been a shortage of broadly-based empirical work in the field. Since theory must be based on empirical findings, and/or be tested empirically, the paucity of survey-based work severely impinges upon progress in the field. A third major hindrance has been a lack of effort to integrate manufacturing strategy ideas with established concepts and theory developed in related disciplines such as business strategy, organization theory, and industrial organization economics. Such integration could yield immediate results in defining environmental and organizational constructs that are presently lacking.

Of course, these three basic causes of the lack of progress are interrelated. For example, theory building often depends on broadly-based empirical work for grounding. Similarly, empirical work itself must be based on constructs developed by theorists in the field or borrowed from other fields. Despite the interrelated nature of the three underlying causes of lack of progress in the field of manufacturing strategy, it is useful to consider them separately. Some detailed observations and suggestions about each of the three are offered below.

Theory Building

One element notably lacking in the published research on manufacturing strategy is a sense of building upon previous research. References to others' work are, for the most part, sparse and casual relative to other fields. Swamidass and Newell (1987) is a notable exception among published work. Progress as a field depends upon the cumulative efforts of the research community as opposed to the highly individual efforts that have thus far characterized the field. The following research suggestions are intended to facilitate theory building.

- **Meta-analysis of published studies in the field.** Such an analysis would include a methodological profile of existing studies and meta-analytic statistical techniques for assessing their empirical findings. Meta-analyses will allow researchers to determine which relationships appear to be solidly established and which areas require more study. This is particularly important since the preponderance of empirical research has thus far focused on a single area: the content of manufacturing strategy.

- **Construct development.** As noted, writers in the field of manufacturing strategy have been casual about establishing their work in the context of what has been previously written. A large part of the price paid for this lack of scholarship in the field is a general inability of scholars in the field to communicate their ideas adequately. Even a cursory review of literature quickly reveals that different authors often discuss the same underlying construct but using different terminology. In order for the field to progress research must be done to identify major constructs. Swamidass and Newell (1987) is an example of research in which major constructs were carefully identified and operationalized.

- **Process research.** Our review of the literature suggests that manufacturing strategy process has been neglected relative to content in both empirical and conceptual work. Adam and Swamidass (1989) recommend that separate research streams be pursued for strategy content and process. This approach would lead to a greater emphasis on process and should lead to theory building in this area.

- **Performance.** Although improved performance is widely acclaimed as the raison d'etre for manufacturing strategy, little concrete work has been done in defining and operationalizing performance measures appropriate for assessing the impact of manufacturing strategy. Two recent studies on performance are promising. Roth (1989) offers an empirical study of performance and strategy and Dixon et al. (1989) provide a conceptual, case-oriented approach. More research on performance and its relationship to the process and content of manufacturing strategy is clearly needed.

- **Interactions among content areas.** Much of the research in manufacturing strategy focuses on a single content area such as quality (Garvin, 1983) or facilities (Schmenner, 1982; 1983). Little research has been aimed at understanding the relationships that exist among content areas or how decisions in one content area affect decisions in other areas. One such effort by Ward (1988) is an empirical study of the relationship between manufacturing technology choice and support staff. Similar studies of interactions among content variables will yield a more powerful model of manufacturing strategy.

- **Interfunctional strategy interactions.** Planned activities (e.g., new product introductions) and capability building progams (e.g., JIT or Total Quality Control) require interactions at the strategic level with other functional areas. The process of how these interactions might be initiated and controlled has been largely ignored in the literature. An exception is Hill's (1989) treatment of the manufacturing-marketing connection. Much more research is required to develop a systemic theory of the relationship among functional strategies.

- **Capability building and manufacturing improvement programs.** A proliferation of programs such as Just-In-Time, Total Quality Control, Computer Integrated Manufacturing, etc., has swept manufacturing practice in recent years. Many of these programs have led to significant improvement in manufacturing capability. Research is needed to establish how these and other capability building programs fit in the process of manufacturing strategy. Miller and Hayslip (1989) begin to address this issue but more research is clearly need.

Empirical Studies

Many critics have noted the relative dearth of survey-based empirical studies in manufacturing strategy (most recently, Adam and Swamidass, 1989). Our review of the literature supports this criticism. Our review also suggests some problems and possible improvements in empirical work in manufacturing strategy.

- **Cross sectional bias.** All of the empirical studies of manufacturing strategy have been cross-sectional. Modeling the cause and effect structure of strategy formulation, implementation, and results requires examination of longitudinal data. Since the business of strategy research often revolves around establishing which actions **cause** success, longitudinal studies would be extremely uselful. Adam and Swamidass (1989) buttress this point by noting the usefulness of PIMS in moving the field of business strategy forward. PIMS data are longitudinal in nature (Schoeffler, Buzzel and Heany, 1974).

- **Data availability.** Manufacturing Futures Project data collected annually in Europe, Japan and North America (Ferdows et al., 1986) are the only ongoing set of survey data. All other data collection projects identified in the literature are "one shot" efforts and relatively small in scale. Based on the PIMS experience in Business Strategy, the availability of single data source to a broad group of researchers would greatly accelerate progress in the field.

- **Unit of Analysis.** Different units of analysis have been used by the various empirical studies (e.g. Miller and Roth (1988) use the "Manufacturing Business Unit", Sharma (1987) uses the factory as the unit of analysis, and Swamidass and Newell (1987) identify the firm as their unit of analysis). Some care is called for in empirical studies to establish the appropriate unit of analysis for a particular research question.

Integration With Other Bodies of Research

In large part, manufacturing strategy has developed independently from related streams of academic research, such as organization theory, industrial organization economics and even business strategy. Recently, research has appeared that does attempt to merge manufacturing strategy with ideas and methods from other disciplines. A notable example of such an effort is Kotha and Orne (1989) who present a conceptual framework linking Porter's (1980, 1985) generic strategies with Hayes and Wheelwright's (1979) product-process life-cycle.

The use of methods, concepts, and constructs developed in other fields promises to greatly speed progress in manufacturing strategy. We present some methods and concepts that are particularly promising for adoption by manufacturing strategy researchers.

- **Methods.** Swamidass and Newell (1987) successfully adopted path analysis from social science research to test a hypothesized model of manufacturing strategy, environment and performance. Because of their willingness to apply an "unusual" methodology these authors were able to use their rather limited data to build theory effectively. Similarly, Sharma (1987) borrowed covariance structure modeling from behavioral science and applied it to his empirical study of manufacturing strategy.

Several Manufacturing Futures Researchers have used various factor analysis and cluster analysis techniques to great advantage.

- **Concepts from business strategy.** Researchers in business strategy have been extremely active in adopting and developing constructs that originated in related disciplines. For example, in their classic contribution to organization theory, Pugh, Hickson, Hinings and Turner (1968) define the important dimensions of organization structure. Business strategy researchers have applied and extended these well tested constructs to great advantage in business strategy (e.g., Miller, 1987).

Similar efforts are required in manufacturing strategy. The payoff is twofold. First, the development of unifying theory will be accelerated by adopting ideas and instruments which have already been tested. Second, researchers in other fields are more likely to better understand manufacturing and manufacturing strategy and further contribute to progress.

CONCLUSION

The most surprising finding is the high level of agreement among researchers as to what constitutes the important issues in manufacturing strategy. This general agreement suggests that many of the major issues have been articulated. However, the literature has not progressed beyond articulating issues and into defining constructs for the purpose of rigorously analyzing these issues. Three causes for this lack of progress have been identified: 1) a paucity of theory construction, 2) little empirical research, and 3) insufficient efforts in adopting ideas and methods from related fields.

We have suggested means for addressing each of these causes. In general, our suggestions involve raising the level of scholarship in manufacturing strategy research while maintaining the practitioner orientation which has been reflected in prior research. We believe that manufacturing strategy research can achieve the high level of scholarship which has been attained, for example, in the field of business strategy.

REFERENCES

Adam, E. E. Jr. and Swamidass, P. M., "Assessing Operations Management from a Strategic Perspective," *Journal of Management*, 15, 2 (1989), 181-203.

Anderson, J. C., Cleveland, G., and Schroeder, R. G., "Operations Strategy: A Literature Review," *Journal of Operations Management*, 8, 2 (1989), 133-158.

Banks, R. L. and Wheelwright, S. C., "Operations vs. Strategy: Trading Tomorrow for Today," *Harvard Business Review*, May-June 1979, 112-120.

Buffa, E. S., *Meeting the Competitive Challenge*, Dow Jones-Irwin, 1984.

Chase, R. B. and Garvin, D. A., "The Service Factory," *Harvard Business Review*, July-August 1989, 61-69.

De Meyer, A., Nakane, J., Miller, J. and Ferdows, K., "Flexibility: The Next Competitive Battle," *Strategic Management Journal*, 10 (1989), 135-144.

Dixon, J. R., Nanni, A. J., and Vollmann, T. E., *Breaking the Barriers: Measuring Performance for World Class Operations*, preliminary edition, Boston: Boston University, 1989.

Fahey, L. and Christiansen, H.K., "Evaluating the Research of Strategy Content," *Journal of Management*, 12 (1986), 167-183.

Ferdows, K., Miller, J. G., Nakane, J. and Vollmann, T. E., "Evolving Global Manufacturing Strategies: Projections into the 1990s," *International Journal of Operations and Production Management*, 6, 4 (1986), 6-16.

Fine, C. H. and Hax, A. C., "Manufacturing Strategy: A Methodology and an Illustration," *Interfaces*, 15, 6 (November-December 1985), 28-46.

Garvin, A., "Quality on the Line," *Harvard Business Review*, September-October 1983, 65-73.

Hayes, R. H., "Strategic Planning - Forward in Reverse?" *Harvard Business Review*, November-December 1985, 111-119.

Hayes, R. H. and Wheelwright, S. C., "Link Manufacturing Process and Product Life Cycles," *Harvard Business Review*, January-February 1979, 133-140.

Hayes, R. H. and Wheelwright, S. C., *Restoring Our Competitive Edge: Competing Through Manufacturing*, John Wiley & Sons, New York, NY, 1984.

Hayes, R. H., Wheelwright, S. C. and Clark, K., *Dynamic Manufacturing*, The Free Press, New York, 1988.

Hill, T. J., *Manufacturing Strategy: The Strategic Management of the Manufacturing Function*, Basingstoke: Macmillan, London, 1985.

Hill, T. J, *Manufacturing Strategy: Text and Cases*, Irwin, Homewood, IL., 1989.

Huete, L. and Roth, A. V., "Linking Manufacturing Capabilities with SBU Strategic Directions," *1987 Proceedings of the Decision Sciences Institute*, Boston, MA, November 1987, 765-767.

Huff, A. S. and Reger, R. K., "A Review of Strategic Process Research," *Journal of Management*, 13, 2 (1987), 211-236.

Johnson, H. T. and Kaplan, R. S., *Relevance Lost*, Boston, MA, Harvard Business School Press, 1987.

Kotha, S. and Orne, D., "Generic Manufacturing Strategies: A Conceptual Synthesis," *Strategic Management Journal*, 10 (1989), 211-231.

Miller, D., "Strategy Making and Structure: Analysis and Implications for Performance," *Academy of Management Journal*, 30, 1 (1987), 7-32.

Miller, J. G. and Hayslip, W., "Implementing Manufacturing Strategic Planning," *Planning Review*, forthcoming, 1989.

Miller, J. G. and Roth, A. V., "Manufacturing Strategies: Executive Summary of the 1987 North American Manufacturing Futures Survey," *Operations Management Review*, 6, 1 (1988), 8-20.

Miller, S. S., "Make Your Plant Manager's Job Manageable," *Harvard Business Review*, January-February 1983, 68-74.

Porter, M. E., *Competitive Strategy*, The Free Press, New York, NY, 1980.

Porter, M. E., *Competitive Advantage*, The Free Press, New York, NY, 1985.

Pugh, D. S., Hickson, D. J., Hinings, C. R., and Turner, C., "Dimensions of Organization Structure," *Administrative Science Quarterly*, 13 (1968), 65-91.

Roth, A. V., "Linking Manufacturing Strategy and Performance: An Empirical Investigation," Working Paper, Boston University, 1989.

Schmenner, R. W., *Making Business Location Decisions*, Prentice-Hall, Engelwood Cliffs, NJ, 1982.

Schmenner, R. W., "Every Factory Has a Life Cycle," *Harvard Business Review*, March-April 1983, 121-129.

Schoeffler, S., Buzzell, R. D. and Heany, D. F., "Impact of Strategic Planning on Profit Performance," *Harvard Business Review*, March-April 1974, 137-145.

Schroeder, R. G., Anderson, J. and Cleveland, G., "The Content of Manufacturing Strategy: An Empirical Study," *Journal of Operations Management*, 6, 4 (1986), 405-415.

Sharma, D., "Manufacturing Strategy: An Empirical Analysis", Unpublished Ph.D. Dissertation, College of Business, The Ohio State University, 1987.

Skinner, W., "Manufacturing - Missing Link in Corporate Strategy," *Harvard Business Review*, May-June 1969, 136-145.

Skinner, W., "The Focused Factory," *Harvard Business Review*, May-June 1974, 112-121.

Skinner, W., *Manufacturing in the Corporate Strategy*, Wiley, New York, NY, 1978.

Skinner, W., *Manufacturing: The Formidable Competitive Weapon*, John Wiley & Sons, New York, NY, 1985.

St. John, C., "Operations Strategy: A Review and Bibliography," *Operations Management Review*, 4 (Summer 1986), 26-32.

Swamidass, P. M., "Manufacturing Strategy: Its Assessment and Practice," *Journal of Operations Management*, 6, 4 (August 1986), 471-484.

Swamidass, P. M. and Newell, W. T., "Manufacturing Strategy, Environmental Uncertainty and Performance: A Path Analytic Model," *Management Science*, 33, 4 (April 1987), 509-524.

Van Dierdonck, R. and Miller, J. G., "Designing Production Planning and Control Systems," *Journal of Operations Management*, 1, 1 (1980), 37-46.

Ward, P., "Overhead Staffing and Process Technology in Manufacturing Industries," Unpublished Doctoral Dissertation, Boston University, 1988.

Ward, P., Miller, J. G., Vollmann, T., "Mapping Manufacturers Concerns and Action Plans," *International J. of Operations and Production Management*, 8, 6 (1988), 5-17.

Wheelwright, S. C., "Reflecting Corporate Strategy in Manufacturing Decisions," *Business Horizons*, February 1978, 57-66.

Wheelwright, S. C., "Japan - Where Operations Really are Strategic," *Harvard Business Review*, July-August 1981, 67-74.

Wheelwright, S. C., "Strategy, Management, and Strategic Planning Approaches," *Interfaces*, January-February 1984, 19-33.

20

THE FOCUS, SPAN AND LINKS IN RESEARCH ON OPERATIONS MANAGEMENT STRATEGY

John P. Matthews and Suan Tong Foo

Graduate School of Business
University of Wisconsin - Madison
1155 Observatory Drive
Madison, Wisconsin 53706
(608) 262-7746

ABSTRACT

A review of Operations Management Strategy (OMS) research shows that a spectrum of issues has already been empirically and theoretically investigated. Collectively, while OMS research efforts have considered many issues, individual research efforts have been more narrowly defined. With the absence of an established framework to place important OMS research, this body of literature takes the appearance of a collection of loosely related efforts. This proposed framework lends additional structure to the field and allows the identification and evaluation of "hot" and "cold" areas of research interest in both management and operations management.

INTRODUCTION

Research in the area of Operations Management Strategy (OMS), while in relative infancy compared with established fields such as corporate and marketing strategy, has still managed to address a wide variety of issues. In a fashion similar to the two blind man who, when patting opposite ends of an elephant, describe two widely different beasts, the individual research efforts in OMS tend to concentrate their focus on limited areas of this large and multifaceted topic. Thus their individual observations, while "locally" accurate and enlightening, may collectively present to others a somewhat confusing mosaic of perceptions and results.

In this paper, we have attempted to accomplish several things. Among them, briefly, our objectives are:

1. To plot and identify areas which have frequently been researched;
2. To identify areas which previously have been only lightly touched or overlooked;
3. To help other researchers be more informed as they continue to explore the true nature of "this beast" called OMS.

One of the first tasks we had to confront was to identify the dimensions along which we would classify research. Two sorts of classifications of articles seemed natural and reasonable:

Literature type: Research appearing in literature devoted to operations management, marketing or management.

Research type: Either empirical or theoretical based research.

Other dimensions of classifications were considerably more problematic since they relate more directly to the actual issues which have been the focus of the research. We decided to concentrate on aspects of research which related directly to things over which the operations manager may have some control or impact. Thus issues such as strategy formulation, consistency, content, implementation and measures of its impact seem reasonable as important elements of this framework. Other important issues generally beyond the control of the operations manager, were noted where appropriate although not employed in the classification. Issues such as structure, levels of authority, span of control and perceived environment uncertainty are examples of issues we noted as having been an element of the research effort but not employed by us as dimensions of classification.

This paper consists of four parts. First, a discussion of terms will be presented. Secondly, a research paradigm consisting of the 5 main topic areas of OMS research will be introduced and examined. Then, a classification of several dozen relevant articles on OMS research, both empirical and theoretical will be reported. Finally, observations on groups of these classifications will be made followed by concluding remarks.

THE MAIN TOPIC AREAS

As a visual aid to this discussion on current literature directions, the five main topic areas to be used in our classification paradigm will be presented as five panels of a regular pentagon (see FIGURE 1). A coding schema was used to classify individual research efforts from both the operations management and the management literature into these five main topic panels (see APPENDIX 1). A review of almost ninety published articles related to OMS revealed that the vast majority of them addressed at least one of the five main topic areas listed in figure 1 below.

The five main research areas have been chosen because they not only reflect the main topics that have received a great deal of attention in the current literature but they also span the entire chain of activities in the firm starting from strategy formulation to the measurement of impact. Also, each of these main topic areas is important to the operations function and can be identified and researched individually as distinct problem areas as well as in combination with other topic areas. These five main areas represent the core of the framework used to classify OMS research. There are situations in which an important part of the research in an article deal with variables other than the five main topic areas. As a service to the reader and in an attempt at completeness, these variables are reflected in the OTHER category (see APPENDIX 1).

FIGURE 1: MAIN TOPIC AREA PENTAGON

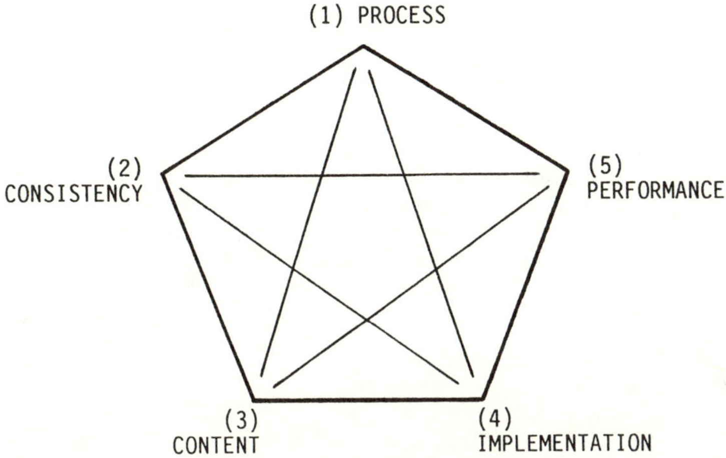

Process

The OMS process area is concerned with the means by which corporate, marketing and manufacturing strategies have been developed. Research in the process area frequently focuses on the role of manufacturing in strategy formulation relative to other functional areas, the timing of entry of manufacturing in these strategy discussions, the level of communications between functional areas, frequency and degree of formality and perhaps other factors relating to strategy development. The research elements for this other areas are listed in APPENDIX 1.

Consistency

Consistency reflects how well the basic elements of strategy at the corporate and functional areas fit together. It addresses the issue of how well-matched the strategies are with respect to the environment, top management, marketing, manufacturing and structure. Consistency does not measure how things have been done but how the strategies fit between and among corporate and functional areas. The typical consistency dimensions that have been used are external and internal consistency with internal consistency having several separate links.

Content

OMS content is concerned with the details of the chosen strategic plan. Content dimensions most frequently used have been quality, price (cost), flexibility and dependability. Content variables, such as service that have not been directly connected to the Hayes and Wheelwright (1984) content dimensions, have been listed under the "other" priorities label. Questions frequently addressed here are the ways in which the firm has competed in its marketplace and the tradeoffs implicitly made in the strategic plan among different aspects of competition.

Implementation

The OMS implementation area addresses the question: "Has the formulated plan been implemented as planned?" Specific elements which appear in OMS implementation research include commitment, resources, time and finances committed to ensure implementation, independence of manufacturing to implement its strategy, and/or culture within the firm toward change.

Performance

This area differs from the others above in the sense that we only look for the presence or absence of performance measures of virtually any kind. We are not comparing research efforts based on the definition and use of performance in the study but rather using a method for classifying types of performance. Performance measure types used are external, internal, single or multi-dimensioned.

SOME KEY CLASSIFICATION PROBLEMS

As with most classification schemes, it is impossible to expect that all the articles in the field of OMS can fit neatly into a single framework. Some of the problems are:

Relative Fit Into Each Category

The problem is that not all articles address the same issue with the same vigor or thoroughness. It is sometimes a judgment call even as to whether an article has a particular feature or not. Placing a relative degree to the presence or absence of a parameter can be even more subjective and therefore is not used here. Because the main objective is to look at the direction of current research and the presence/absence in a particular area, the 0-1 code used is deemed adequate for this purpose.

Structure

To this day, there is no set definition of structure. Many issues (e.g. size, technology) have typically been called structure elements. We tried to follow the current management definitions in using this term but cannot be more precise than what is found in the literature.

Choice Of Articles

There is no doubt that this can be a important facet of this type of research. We tried to have the area research fairly represented without selection bias. Generally, these articles come from sources like Journal of Operations Management (JOM), Decision Science (DS), Decision Sciences Proceedings (DSP), Management Science (MS), Journal of the Operations Research Society (JORS), Interfaces (I), International Journal of Production Research (IJPR), Harvard Business Review (HBR), Journal of Management (JM), Strategic Management Journal (SMJ), Academy of Management Journal (AMJ), Academy of Management Review (AMR), Administrative Science Quarterly (ASQ), Management Review (MR), Long Range Planning (LRP), Columbia Journal of Business (CJB), Sloan Management Review (SMR), California Management Review (CMR), Journal of Marketing (JMK), Journal of Marketing Research (JMR), European Journal of Marketing (EJM), working papers (WP) and books (BK). Although an exhaustive search has been carried out, it is possible that some important articles are missing. The patterns reported here are of course a function of the articles listed. The authors welcome enlightenment from readers for comments on important literature we may have overlooked.

CODING SCHEME AND TABLE EXPLANATION

Table 1 is a count of the frequency of occurrence of each element in the main topic areas. This is ordered by research type, literature, then lastly name. The coding scheme used is based on 0-1 codes where 1 depicts the presence of an element within the key topic area in the article and a 0 (blank) refers to the absence of the element. As an example concerning how to read table 1, we will use the article by Rafii and Miller in the 11th row of this Table in the APPENDIX. Reading across this row, we can see that this article addresses the elements of:

1. Role and Communication in Process (PROCESS A AND B),
2. Corporate/Manufacturing and Manufacturing/Structure Consistency (CONSISTENCY B AND E),
3. Multiple internal measures (PERFORMANCE B AND C),
4. Structure (OTHER A),
5. Empirical data source is a data base (DATA C),
6. Operations management strategy literature (LITERATURE A),

OBSERVATIONS

The observations concerning the direction of the strategy research can be made along two dimensions, representing different levels of observational detail, These dimensions are:

1) General overall observations concerning OMS research
2) Observations within each key topic area and the relationship between and among key topic areas

General Overall Observations Concerning OMS Research

From the general lists of articles (TABLE 1: ARTICLE LIST AND SUMMARY COUNT DATA), it can be seen that there is generally a greater amount of theoretical research than empirical research. Additionally, empirical research typically has a more simple model structure, focusing on a single key topic area or combining a few of the key areas together one integrated model. The reason for this narrowness of focus is probably obvious to anyone contemplating empirical research. The problems of manageability, parsimony and measurability can be formidable. It seems that there is more theory postulated than empirically verified.

The most researched topics in empirical OMS have been Content with Process being a distant second and Consistency not far behind. The least addressed issue has been Implementation. For the theoretical research in OMS, Process and Consistency have been most frequently addressed, with Content being addressed less often. Similarly, Implementation has commanded the least attention. The emphasis in Content research in empirical OMS is probably a function of the availability of information. The lack of measures and contructs in Implementation may have hindered the development of Implementation in empirical work. This has also been a problem for the management empiricists although the issues have recently been the focus of more attention (Nutt 1987,1989).

Typically, only empirical work addresses the performance problem so our discussion on performance will be limited to this work. It is interesting that the majority of research uses only internal measures to evaluate firm performance, although external measures may be more important particularly when the motivation for change of strategy is external. External measures have been used only infrequently. Some multiple measures have been used reflecting the multi-dimensional nature of firm effectiveness.

Observations Within Each Topic Area And Relationship Between And Among Different Key Topic Areas

This section will detail the characteristics of each key topic area as well as comparing and contrasting the links between each of the key topic areas of research.

Process. Within the OMS research, process has not been a frequent topic of study. This may be because a behavioral topic area like process has not been traditionally emphasized in operations management. Additionally, there does not seem to be a strong pattern reflecting the tendencies of researchers to include or exclude process when studying other functional areas in either the OMS or management research literature. The same pattern or lack thereof can be found in both the empirical and the theoretical research.

Consistency. Consistency has been a much more frequently studied topic in theoretical than in empirical research. Also, theoretical research considers more elements of Consistency while empirical research considers at most two. Consistency in corporate/manufacturing strategy is most often studied in empirical research, while all types of Consistency are evenly represented in theoretical research. This probably reflects the difficulty in definition and measurement of this concept. There have been many models and theories linking Consistency to firm effectiveness but whether these concepts actually have significant influence on firm effectiveness has not been tested in the OMS empirical literature. In looking at the relationship of Consistency to other key topic areas, no strong patterns can be observed in either operations management or management literature. This same lack of a discernible pattern in research links between consistency and other areas exists in both the empirical and theoretical research.

Content. Content has been one of the most frequently addressed issues in empirical operations management research. Most research in this area has been consistent in the use of the Hayes and Wheelwright (1984) definition of content (cost, dependability, quality, flexibility). Theoretical literature has also exhibited a good deal of interest in Content. The management literature when taken as a group, reflects a great deal of Content research but because of the definitional differences between the established OMS conventions of this paper and management conventions, it cannot be reflected in the classification system employed here. Many papers in management do address the issues they regard as Content. However, what OMS regards as Content is different from what the management regards as Content. In empirical OMS research, the Content area has most often been studied by itself, separate from other areas. The Content theoretical research tends to link Content with the areas of Process and Consistency.

Implementation. Implementation research has definitely been under-represented in both OMS and management research. With the exception of management information systems and operations research, very little has been written about Implementation models and strategies. Because one cannot assume that all planned strategies are implemented well, Implementation research becomes important to strategy evaluation and performance. Since little has been written about Implementation in the empirical OMS literature, it is difficult to make statements about the direction of research. In both empirical and theoretical management research, Implementation has been more frequently addressed than in OMS. When Implementation has been addressed in this literature survey, there has been a tendency to link it with Consistency. The same pattern has been exhibited in both the empirical and theoretical models.

Performance. The method of evaluation of the impact of strategy particularly at the operations management level has been a contested issue. We will not fuel the debate regarding which type of performance measure is most appropriate However, as noted earlier, single internal measures has been most commonly used with multiple internal measures not as common. We note that there has been very little research employing external measures (single or multiple) or a combination of external and internal measures. This may be a function of data availability. Finally, we observe that content research tends to be descriptive and does not evaluate performance.

THOUGHTS ON FUTURE DIRECTIONS

A frequently voiced axiom regarding research is that empirical research tends to follow theory development. A second axiom is that the management literature lends OMS literature in the discussion of issues of common interest. We will not be testing these axioms but be using them for purposes of projecting trends in the current research. In the first axiom is true, then Consistency can be expected to emerge as an area of more intensive empirical research interest in the OMS literature since its popularity in the OMS theoretical literature is much more pronounced (see FIGURE 2 below).

FIGURE 2: SUMMARY OF RESEARCH EFFORTS

LITERATURE	TYPE	INTENSITY OF RESEARCH: HOT	MODERATE	COLD
OMS	EMPIRICAL	CONTENT	PROCESS CONSISTENCY PERFORMANCE	IMPLEMENTATION
OMS	THEORETICAL	CONSISTENCY CONTENT	PROCESS	IMPLEMENTATION
MGT	EMPIRICAL	CONSISTENCY PERFORMANCE	IMPLEMENTATION PROCESS	
MGT	THEORETICAL	CONSISTENCY	IMPLEMENTATION PROCESS	

If the second axiom is true, since implementation has been a relatively common aspect of both theoretical and empirical research in management when compared to OMS, perhaps it will soon be gaining greater attention in the OMS literature (see FIGURE 2 above). At present, implementation is a topic that is rarely addressed in both the OMS empirical and theoretical literature. When looking at all the articles cited in this study as a group, with respect to the relative recency of publication, one can see that there has been a spate of empirical OMS articles of recent origin which has specifically addressed Content. Only one article of the eleven cited in the OMS literature treating the content area has been published before 1986. The authors feel that while this interest has been intense, it is reasonable to expect that interest cannot long remain at this high level and that other areas will receive more attention in the future.

Finally, we note that there has not been a strong research trend toward the use of increasingly complex models in the OMS empirical literature. Frankly, this surprised us. We have expected to observe that over time, more issues would be addressed and empirically tested in OMS models. This was not found to be the true in the cited articles. Perhaps the difficulty in measuring many of the variables (which we call elements in each of the topic areas) which describe the areas of interest mitigate against more broad and inclusive models.

APPENDIX 1: LITERATURE SUMMARY EXPLANATION (FOR TABLE 1)

OMS PROCESS
This describes the process of strategy formulation in the firm
- (A) ROLE OF MANUFACTURING IN STRATEGY FORMULATION
- (B) COMMUNICATION DURING STRATEGY FORMULATION
- (C) OTHER

CONSISTENCY
This describes how well the firm is able to get a consistent direction for top management and other functional areas
- (A) EXTERNAL
- (B) INTERNAL - CORPORATE / MANUFACTURING
- (C) INTERNAL - MARKETING / MANUFACTURING
- (D) INTERNAL - MANUFACTURING
- (E) INTERNAL - MANUFACTURING / STRUCTURE
- (F) OTHER

CONTENT
This describes the priorities used in planning (actual plans)
- (A) COST
- (B) QUALITY
- (C) FLEXIBILITY
- (D) DEPENDABILITY
- (F) OTHER

IMPLEMENTATION
This describes the process of implementing the firm' strategy
- (A) COMMITMENT
- (B) INDEPENDENCE
- (C) CULTURE
- (D) OTHER

PERFORMANCE
This describes the effectiveness measures used in the study
- (A) SINGLE MEASURE
- (B) MULTIPLE MEASURES
- (C) INTERNAL MEASURE
- (D) EXTERNAL MEASURE
- (E) UNSPECIFIED

OTHER
These are variables of significance in the reviewed papers but are not directly addressed in this study
- (A) STRUCTURE/TECHNOLOGY
- (B) ENVIRONMENT (PEU)
- (C) CONTROL
- (D) INTERNATIONAL
- (E) COMPETENCE

DATA SOURCE/TYPE
This describes the data sources used in the reviewed papers
- (A) QUESTIONNAIRE/EMPIRICAL
- (B) INTERVIEWS/EMPIRICAL
- (C) DATA BASE/EMPIRICAL
- (D) THEORETICAL

LITERATURE TYPE
This describes the main topic affiliation of the paper
- (A) OPERATIONS MANAGEMENT STRATEGY
- (B) MANAGEMENT
- (C) MARKETING

CODING SCHEME (FOR EACH OF THE ELEMENTS ABOVE)
- 1 = PRESENCE
- BLANK = ABSENCE

TABLE 1: ARTICLE LIST AND SUMMARY COUNT DATA
ORDERED BY LITERATURE (PRIMARY)
NAME (SECONDARY/DESCENDING ORDER)

			PROC			CONSISTENCY						CONTENT					IMPL(1)				PERF				OTHER					DATA				LIT		
AUTHOR	REF	DATE	A	B	C	A	B	C	D	E	F	A	B	C	D	E	A	B	C	D	A	B	C	D	A	B	C	D	E	A	B	C	D	A	B	C
CLEVELAND ET AL	WP	88														1						1	1		1				1		1			1		
DEMEYER ET AL	SMJ	89										1	1	1	1	1									1			1				1		1		
DEMEYER,FERDOWS	IJPR	87										1	1	1	1					1							1					1		1		
ETTLIE,REIFEIS	I	87			1															1					1						1			1		
FERDOWS,LINDBERG	IJPR	87										1	1	1	1	1									1							1		1		
HAYES,CLARK	I	85																				1	1		1						1			1		
HORTE ET AL	IJPR	87										1	1	1	1										1							1		1		
HEUTE,ROTH	DSP	87										1	1	1	1																	1		1		
MCDOUGALL	D	86		1																	1		1		1					1				1		
MEREDITH	I	87																		1					1		1				1			1		
RAFII,MILLER	WP	83	1	1			1			1												1	1		1							1		1		
RICHARDSON ET AL	I	85					1		1												1		1		1					1				1		
SCHMENNER,COOK	JOM	85																				1	1	1	1					1				1		
SCHROEDER ET AL	JOM	86			1		1		1			1	1			1													1	1				1		
SWAMIDASS	WP	86					1					1	1	1	1															1				1		
SWAMIDASS,NEWELL	MS	87	1									1	1	1	1							1	1			1				1				1		
VAN DIERDONCK ET AL	JOM	83			1											1					1		1		1	1					1			1		
YONSEI UNIVERSITY	WP	88										1	1	1	1	1									1		1					1		1		
COOL,SCHENDEL	SMJ	88						1			1										1		1		1	1				1					1	
DEWAR,DUTTON	MS	86														1				1	1		1							1					1	
GIBSON	BK	76															1	1	1	1	1		1		1					1					1	
GOVINDARAJAN	SMJ	89									1									1		1	1		1		1			1					1	
GOVINDARAJAN	AMJ	88																1		1		1	1				1			1					1	
HATFIELD,HUSEMAN	AMJ	82		1			1			1											1		1		1					1					1	
HITT,IRELAND	DS	84																			1			1	1	1			1	1					1	
LUCAS	BK	76																		1	1		1							1					1	
MAHAJAN,WIND	LRP	88					1		1	1											1		1		1							1			1	
MCMAHON,IVANCEVICH	ASQ	76		1	1					1											1		1		1		1			1					1	
NUTT	SMJ	89															1			1	1		1		1						1				1	
NUTT	SMJ	87															1			1	1		1								1				1	
SCHULTZ,SLEVIN	BK	76		1	1														1	1	1		1							1					1	
SORENSEN,ZAND	BK	76			1															1		1	1							1					1	
VAN DE VEN,WALKER	ASQ	84		1	1					1															1					1					1	
VERTINSKY ET AL	BK	76																	1	1	1		1							1	1				1	
TOTAL			2	6	7	0	6	1	3	5	2	9	9	8	8	7	3	2	3	13	15	8	22	2	22	4	6	1	3	19	8	8	0	18	16	0

			PROC			CONSISTENCY						CONTENT					IMPL(1)				PERF				OTHER					DATA				LIT		
AUTHOR	REF	DATE	A	B	C	A	B	C	D	E	F	A	B	C	D	E	A	B	C	D	A	B	C	D	A	B	C	D	E	A	B	C	D	A	B	C
ABERNATHY	DS	76			1					1															1		1						1	1		
ADAM,SWAMIDASS	WP	88	1	1	1							1	1	1	1	1												1					1	1		
FINE,HAX	I	85			1		1	1	1																1								1	1		
GUDNASON,RIIS	O	84			1																				1								1	1		
HAX,MAJLUF	I	88	1	1	1	1	1		1	1																							1	1		
HAYES,SCHMENNER	HBR	79			1		1																		1		1		1				1	1		

TABLE 1: ARTICLE LIST AND SUMMARY COUNT DATA (CONTINUED)
ORDERED BY LITERATURE (PRIMARY)
NAME (SECONDARY/DESCENDING ORDER)

AUTHOR	REF	DATE	PROC			CONSISTENCY						CONTENT					IMPL(1)				PERF				OTHER					DATA				LIT		
			A	B	C	A	B	C	D	E	F	A	B	C	D	E	A	B	C	D	A	B	C	D	A	B	C	D	E	A	B	C	D	A	B	C
HAYES,WHEELWRIGHT	HBR	79						1																	1								1	1		
HAYES,WHEELWRIGHT	HBR	79						1																	1								1	1		
HILL	BK	85	1	1	1		1	1	1			1	1	1	1										1				1				1	1		
HOBBS,HEANEY	HBR	77			1		1			1																							1	1		
KOTHA,ORNE	SMJ	89						1																	1		1		1				1	1		
MILLER	HBR	81								1															1								1	1		
PENDLEBURY	LPR	88			1											1				1					1								1	1		
SCHROEDER	MR	84			1		1		1			1	1	1	1																		1	1		
SHAPIRO	HBR	77						1																									1	1		
SKINNER	BK	85	1	1	1	1	1	1		1										1					1				1				1	1		
SKINNER	HBR	74	1	1	1					1						1																	1	1		
STOBAUGH,TELESIO	HBR	83														1									1								1	1		
SWAMIDASS	JOM	86												1											1								1	1		
WHEELWRIGHT	I	84	1									1	1	1	1																		1	1		
ALEXANDER	LPR	85															1	1	1	1					1		1						1		1	
AROGVASAWY,BYLES	JOM	87					1			1															1								1		1	
BAIRD,MESHOULAM	AMR	88				1					1									1													1		1	
BARNEY	MS	86																		1					1	1	1						1		1	
BLOIS	IMM	85						1																	1								1			1
BOURGEOIS,BRODWIN	SMJ	84																		1					1		1						1		1	
CHAKRAVARTHY ET AL	I	84			1						1									1													1		1	
CHURCHMAN	BK	76			1															1													1		1	
COURETAS	BM	85						1																	1								1			1
DESS,ORIGER	AMR	87			1		1																		1	1							1		1	
DRAZIN,VAN DE VEN	ASQ	85				1		1		1															1								1		1	
DUFFY	SMJ	89								1											1		1				1						1		1	
FRY,SMITH	AMR	87								1															1								1		1	
GUMMESSON	LRP	87				1		1																					1				1			1
GUPTA	AMR	84					1																		1								1		1	
HUSSEY	LRP	85																		1					1								1		1	
JELINEK,BURNSTEIN	AMR	82			1	1		1		1	1														1								1		1	
JELINEK,GOLDHAR	CJB	83																							1		1						1		1	
KAZANJIAN,DRAZIN	AMR	87						1												1													1		1	
LEONTIADES	JBR	83				1		1																	1				1				1			1
MAHER	BM	84						1																									1			1
MCEWEN ET AL	LRP	88									1						1	1	1	1													1		1	
MEHRA,BRETZ	I	81			1															1													1		1	
MEYER	EJM	83			1			1																	1								1		1	
MILES,SNOW	CMR	84				1				1																							1		1	
MILES,SNOW	CMR	86				1	1																										1		1	
NEMETZ,FRY	AMR	88			1					1															1				1				1		1	
QUINN	SMR	80		1	1															1													1		1	
REED,BUCKLEY	LRP	88								1										1		1		1	1		1						1		1	
SCHOLZ	LRP	87				1					1														1								1		1	
SCHULTZ,SLEVIN	BK	76		1	1															1													1		1	
TOTAL			6	7	21	10	11	16	4	14	5	4	4	5	4	4	2	2	2	15	1	1	1	1	31	2	9	1	7	0	0	0	51	20	26	5

COMPLETE REFERENCES AVALIABLE UPON REQUEST

21

INVENTORY REDUCTION AND PRODUCTIVITY GROWTH: A STUDY OF JAPANESE AUTOMOBILE PRODUCERS

Marvin B. Lieberman

Hoover Institution
Stanford, CA 94131
(415) 725-8557

ABSTRACT

This paper assesses trends in WIP inventory levels and labor productivity for a sample of eight Japanese motor vehicle manufacturers. Inventory buffers held between component suppliers and final assembly firms are also evaluated. In addition, a methodology is proposed for measuring materials throughput time on a consistent basis.

The results show a strong connection between materials throughput time and labor productivity, particularly during the 1960's and early 1970's. By the 1980's, most of the Japanese producers had adopted "just-in-time" production methods, leading to greater similarity in throughput time and a weakening of the connection between throughput time and productivity. During the same period, closer coupling was achieved between final producers and component suppliers, as intermediate parts inventories were reduced.

INTRODUCTION

This paper reports the early findings of a study of inventory reduction and productivity growth among Japanese motor vehicle manufacturers. The sample includes eight firms—Toyota, Nissan, Mazda, Daihatsu, Isuzu, Fuji (Subaru), Suzuki and Honda—over a period from the early 1960's to 1987. The study makes use of data on raw material and work-in-process inventories that are reported by all major Japanese manufacturing companies. (Inventory data reported in the US are much less detailed.)

Findings of Prior Study

This research extends the findings of a prior study of comparative productivity among US and Japanese automotive producers (Lieberman, Lau and Williams, 1989; hereafter referred to as LLW). LLW compared the productivity of six major US and Japanese motor vehicle manufacturers (General Motors, Ford, Chrysler, Toyota, Nissan and Mazda) from the early 1950's through 1987. Three potential determinants of growth in total factor productivity were evaluated: economies of scale, adoption of "just-in-time" manufacturing systems, and changes in company top management.

The LLW study found that productivity improvement by all six motor vehicle producers was attained primarily through more efficient utilization of labor; long-term growth in capital productivity was negligible for most firms. The three Japanese producers began at low productivity levels but achieved higher labor productivity than their US counterparts by the late 1970's. During the 1980's, however, productivity grew rapidly at Ford and Chrysler while some of the Japanese firms made only minor gains. As a consequence, productivity differences among firms within each country now appear large relative to the gap in average productivity between the US and Japan.

Early productivity growth for Japanese producers was derived in part from the attainment of scale economies, but this source of improvement was largely exhausted by the mid-1960's. Top management was found to have had a major impact on company productivity. For all producers except Toyota, signficant shifts in the level and growth rate of productivity followed changes in top management.

The LLW findings suggest that productivity growth stemmed in part from the implementation of just-in-time manufacturing systems. Toyota, the pioneer in developing these systems, experienced the most rapid productivity growth over the sample period. Mazda also had a dramatic improvement in productivity when it adopted key elements of the Toyota system. Reduction in work-in-process (WIP) inventories is an important component of just-in-time manufacturing. Producers in both countries made substantial inventory reductions, and periods of rapid growth in company productivity often coincided with periods of inventory reduction. However, statistical tests failed to support the significance of inventory reduction when controls were included for other plausible sources of productivity growth.

METHODOLOGY

This paper reports basic trends in work-in-process inventory levels and labor productivity for the sample of eight Japanese motor vehicle manufacturers. Parts inventories held between component suppliers and the final assembly firms are also evaluated. A subsequent paper will provide more detailed statistical tests of the relation between inventory levels and productivity.

Labor productivity is measured as value-added per employee per year, in constant 1985 Yen. This is similar to the hourly productivity measure reported in LLW. The method for computing labor productivity from financial data in company annual reports is described in LLW.

Reduction in work-in-process inventory is an essential feature of any "just-in-time" manufacturing program (c.f., Schonberger, 1982; Hall, 1985). Inventory levels can be characterized using a variety of alternative measures: e.g., WIP "turns" per year, WIP as a percent of sales, or WIP throughput time. In comparing WIP levels across companies, it is necessary to adjust for differences in vertical integration. This study computes adjusted measures of WIP throughput time, as described below.

Consider, first, a fully-integrated firm (i.e, a firm where value-added equals 100% of sales). Assume that work-in-process moves through sequential processing stages and intermediate buffer inventories at a constant speed. Further, assume that as work-in-process flows through the processing stages, it accumulates accounting value in a linear fashion. Define the total manufacturing throughput time, T, as the time required for a product to move through the entire manufacturing process, beginning as raw materials input (value=0) and ending as finished output. For analytical simplicity, assume that profit margins are zero, so that sales, S, equals the cost of goods sold. Under these assumptions, the amount of work-in-process inventory, W, is:

$$W = \frac{1}{2}ST.$$

Thus, given data on W and S, one can calculate the total manufacturing throughput time, $T = 2W/S$.[1]

This formula must be modified to account for the fact that firms are seldom fully integrated, i.e., value-added (V) is less than sales. For a partially-integrated firm, the value of the firm's work-in-process is:

$$W' = \frac{1}{2}(S + M)T',$$

where M represents the the cost of purchased materials, and T' is the time required for these materials to be transformed to finished goods. T' may be termed the producer throughput time,[2] which will always be less than or equal to the total manufacturing throughput time, since a partially-integrated firm performs only a subset of the operations required to produce the final product.

[1] Alternatively, $1/T$ gives the number of WIP "turns" per year.

[2] Note that $T' = 2W'/(S + M)$.

As an illustrative example, consider a fully-integrated firm with \$100 million annual sales and work-in-process inventory of \$20 million. The total manufacturing throughput time is $20/.5(100) = 0.4$ years. Assume that this firm is split into two vertically-related firms, (i.e., parts supplier and final producer), each with value-added of \$50 million. Moreover, assume that the finished goods inventory held by the parts supplier and the raw materials inventory held by the final producer are negligible. Under these assumptions, the work-in-process inventory of the parts supplier is \$5 million and the work-in-process inventory of the final producer is \$15 million. The throughput time for the parts supplier is $5/.5(50) = 0.2$ years, and the throughput time for the final producer is $15/.5(50 + 100) = 0.2$ years. Hence, the total cycle time remains 0.4 years, as illustrated by the following diagram:

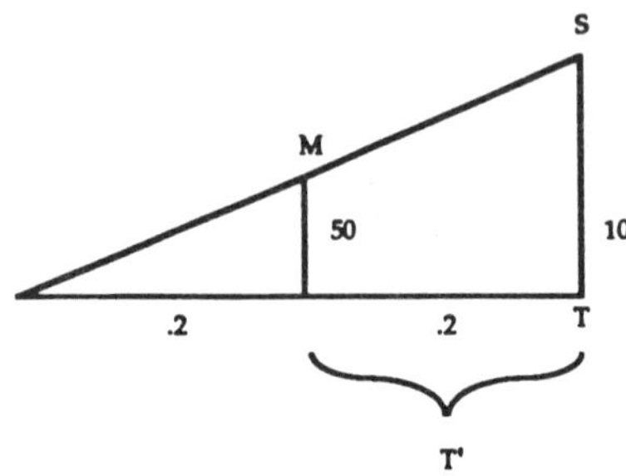

To provide a benchmark for comparing firms that make similar final products but have different degrees of backward integration, a total manufacturing throughput time can be extrapolated from the producer time by the formula:

$$T = T'(S/V).$$

This measure is used to compare the eight Japanese auto producers.

It is also be possible to examine the magnitude of inventory buffer held between parts suppliers and final producers. This buffer consists of (1) finished goods inventory held by parts suppliers, and (2) raw materials inventory held by final producers. For parts suppliers, the appropriate measure is average finished goods inventory divided by total annual sales. For final producers, the appropriate measure is average raw materials inventory divided by annual materials purchases.[3]

THROUGHPUT TIME COMPARISON

The figures on the following page give the materials throughput time of each producer, plotted over the 1964-87 period.[4] The throughput times were estimated from annual data on sales, value added, and WIP inventory,

[3]The latter equals sales minus value added.

[4]Data for Honda are available only for 1975 and later years, when the firm was engaged in automobile production.

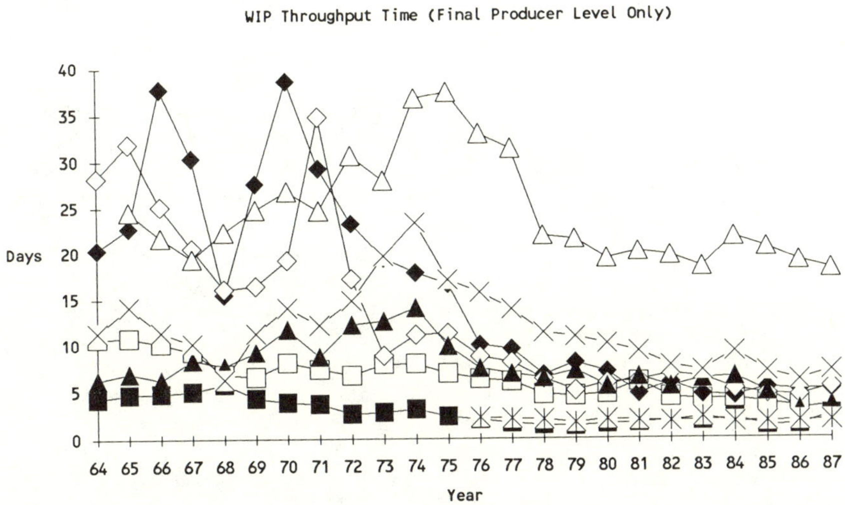
WIP Throughput Time (Final Producer Level Only)
Days
Year
40
35
30
25
20
15
10
5
0
64 65 66 67 68 69 70 71 72 73 74 75 76 77 78 79 80 81 82 83 84 85 86 87

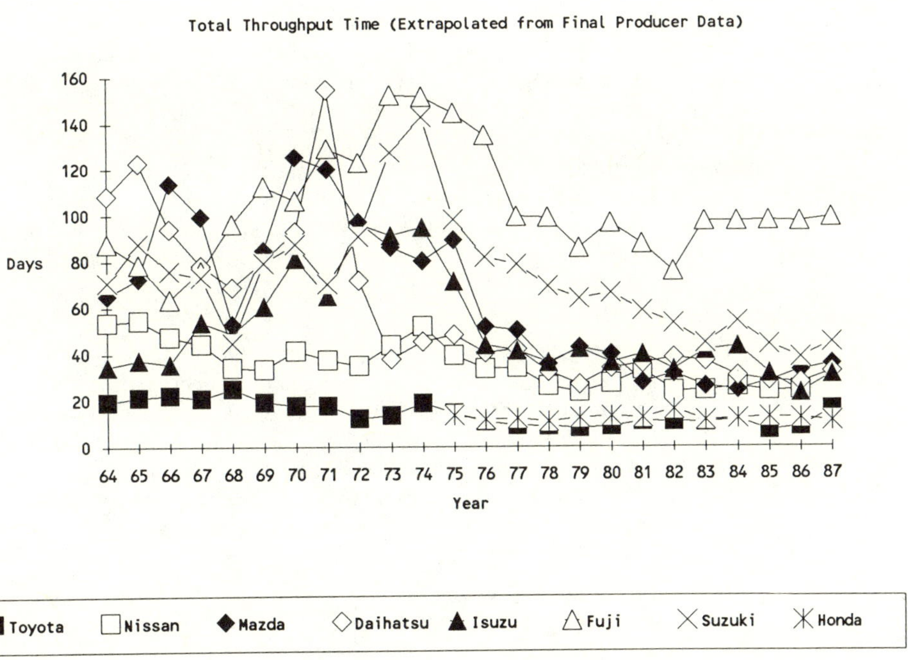
Total Throughput Time (Extrapolated from Final Producer Data)
Days
Year
160
140
120
100
80
60
40
20
0
64 65 66 67 68 69 70 71 72 73 74 75 76 77 78 79 80 81 82 83 84 85 86 87
Toyota
Nissan
Mazda
Daihatsu
Isuzu
Fuji
Suzuki
Honda

using the formulas above. The top graph shows the throughput time of each final producer (T'). The bottom graph adjusts for differences in producer vertical integration. (Other factors equal, throughput time increases with the extent of integration.) The bottom graph gives estimates of the total manufacturing throughput time (T), which includes the estimated throughput time of suppliers under the assumption that suppliers have the same speed of materials flow as the final producer being examined.

The top graph shows that thoughput time has been declining for all producers except Fuji. Toyota had the fastest throughput of all Japanese producers in the 1960's, and the firm has continued to make significant improvements. Honda shows throughput times comparable to those of Toyota in recent years. The data suggest that many of the Japanese firms adopted "just-in-time" systems in the mid-1970's. In particular, Mazda and Daihatsu exhibited high WIP levels and fluctuating throughput times during the 1960's and early 1970's. Subsequently, both firms made dramatic improvements. In the 1960's average throughput times ranged from about 5 days at Toyota to nearly 40 days at Mazda, Daihatsu and Fuji. By the mid-1980's all firms except Fuji had reduced their average throughput time to less than 10 days.

The bottom graph gives an estimate of the total manufacturing throughput time, assuming that supplier firms processed materials at the same speed as the downstream assembler. This measure corrects for differences in vertical integration across the eight producers as well as changes in the vertical integration of any given producer over time. (In general, the final assembly firms were reducing their degree of integration.) The trend toward faster throughput time, and convergence among producers, is less pronounced. Hence, some of the acceleration in throughput time suggested by the upper graph is simply an artifact of firms' reduced scope of operation.

LABOR PRODUCTIVITY

The figure on the following page plots real labor productivity for each of the eight producers. The graph shows that labor productivity grew significantly for all producers over the 1964-87 period. During the 1960's and 1970's, the labor productivity of Toyota and Nissan was considerably higher than that of the other Japanese producers. During the 1980's, however, Nissan's productivity growth stalled while Toyota expanded its lead. In recent years, Honda has been the second most productive manufacturer.[5]

[5] Honda's measured productivity may be depressed somewhat given that a large fraction of the firm's international sales transactions are in dollars.

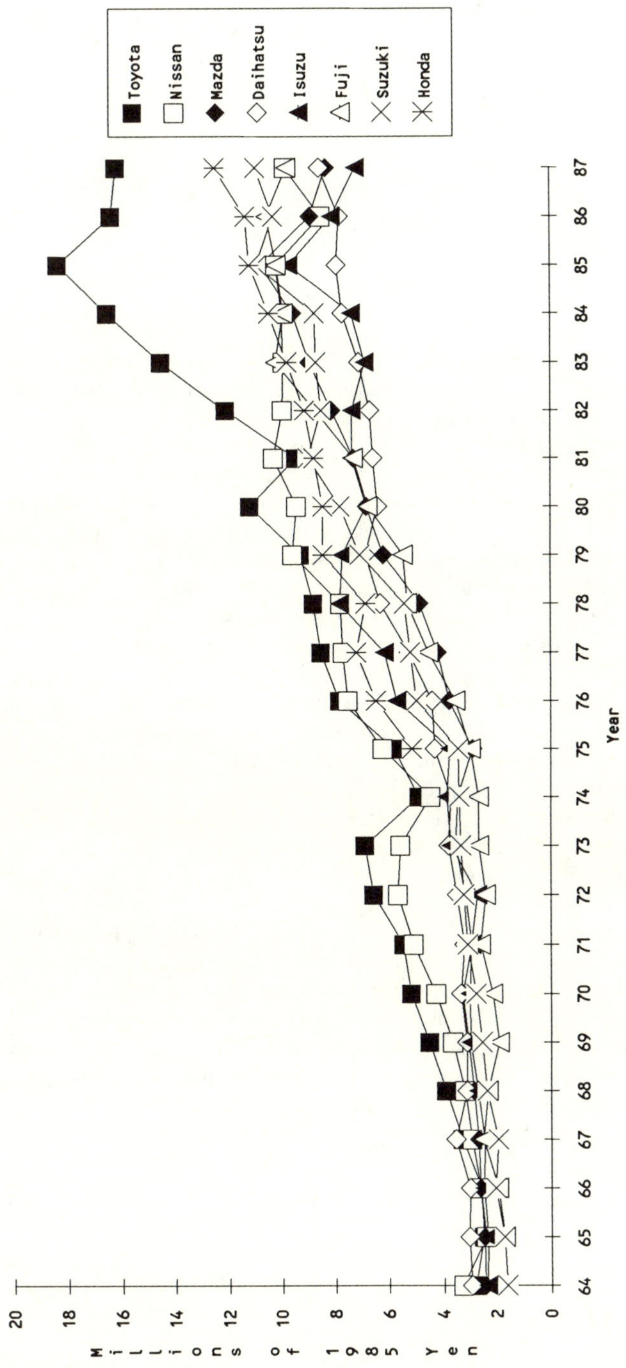
Labor Productivity (Value-Added per Employee per Year)
Toyota
Nissan
Mazda
Daihatsu
Isuzu
Fuji
Suzuki
Honda
Millions of 1985 Yen
0
2
4
6
8
10
12
14
16
18
20
64
65
66
67
68
69
70
71
72
73
74
75
76
77
78
79
80
81
82
83
84
85
86
87
Year

RELATION BETWEEN THROUGHPUT TIME AND LABOR PRODUCTIVITY

The connection between throughput time and productivity is of primary interest. The figures on the following page plot labor productivity versus throughput time for each of the producers in a given year. The top graph is for 1973; the bottom graph is for 1987. These are two representative years, separated by a period of more than a decade. For each year, the vertical axis gives the firm's labor productivity relative to the average labor productivity of all producers in that year. The horizontal axis gives the firm's estimated total manufacturing throughput time (T) relative to the average throughput time of all producers in the observation year.

The top graph documents a remarkably strong connection between throughput time and labor productivity in the early 1970's. Toyota had the fastest materials throughput and was the most productive manufacturer. Nissan had throughput time and productivity close to that of Toyota. Fuji had the longest throughput time and was the least productive; other companies were intermediate. The points for the seven firms lie along a near-linear fit.

The relation in the bottom graph is similar but less pronounced. This suggests that as the variation in throughput time among producers has diminished, differences in labor productivity have become less closely linked to differences in throughput time.

While these results are suggestive, they do not necessarily indicate any causal connection between throughput time and labor productivity. Throughput time and labor productivity may both be determined by changes in external factors. A subsequent study will provide more detailed statistical tests, including controls for other factors. In addition, the labor productivity measure will be supplemented by measures of total factor productivity, which control for changes in capital input.

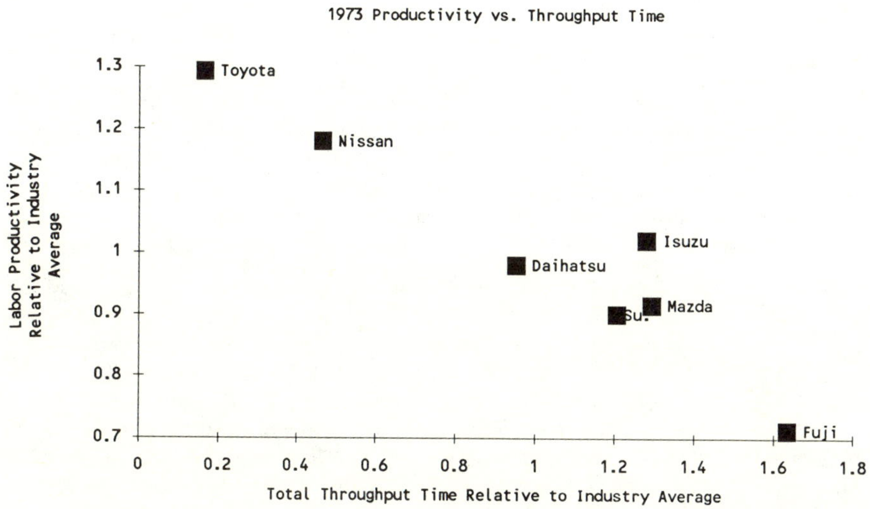
1973 Productivity vs. Throughput Time
Labor Productivity Relative to Industry Average
Total Throughput Time Relative to Industry Average
Toyota
Nissan
Isuzu
Daihatsu
Su.
Mazda
Fuji
1.3
1.2
1.1
1
0.9
0.8
0.7
0
0.2
0.4
0.6
0.8
1
1.2
1.4
1.6
1.8

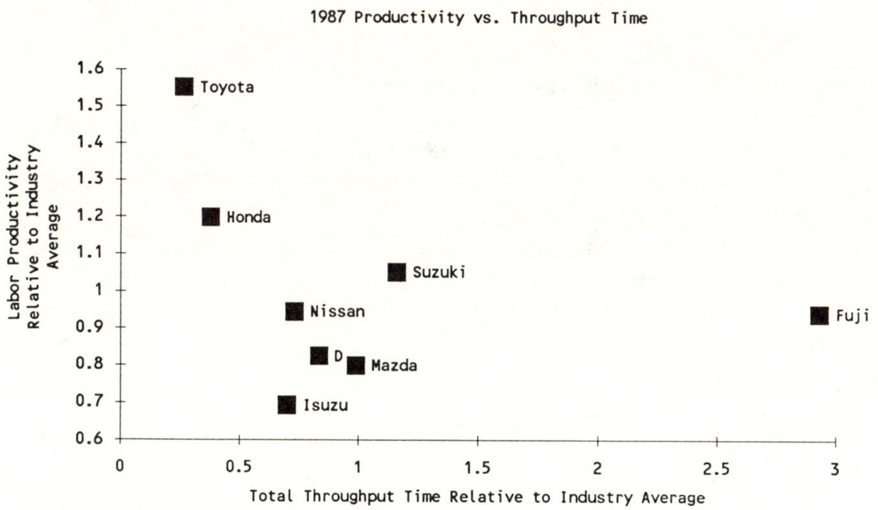
1987 Productivity vs. Throughput Time
Labor Productivity Relative to Industry Average
Total Throughput Time Relative to Industry Average
Toyota
Honda
Suzuki
Nissan
Fuji
D
Mazda
Isuzu
1.6
1.5
1.4
1.3
1.2
1.1
1
0.9
0.8
0.7
0.6
0
0.5
1
1.5
2
2.5
3

INVENTORY BUFFERS BETWEEN COMPONENT SUPPLIERS AND FINAL PRODUCERS

In addition to reductions in WIP inventory held within the firm, "just-in-time" manufacturing requires tighter coupling with suppliers. The graphs on the following page describe changes in the level of buffer inventory held between parts suppliers and final producers. The top figure plots changes in the level of finished goods inventories held by eighteen Japanese automotive parts suppliers between 1976 and 1987. The bottom figure plots changes in the levels of raw materials inventories held by the eight automotive final producers.

Major reductions in inventory levels are apparent for both the parts suppliers and the final producers. In general, firms that held the largest inventories in 1976 tended to make the greatest cuts.[6] All of the final producers reduced their raw materials inventory levels, and all but one of the parts suppliers cut their relative inventories of finished goods. (The one exception, Tokai, had the lowest inventory level in 1976, and one of the lowest in 1987.) These results imply that the coupling between final producers and parts suppliers was becoming tighter over time.

Critics of "just-in-time" manufacturing in the US have sometimes alleged that final producers achieve reductions in raw materials inventories at the expense of upstream suppliers, who are forced to hold larger stocks of finished parts. The data show that, in general, this has not been the case in the Japanese automotive sector.

REFERENCES

Hall, R. W., *Zero Inventories.* Homewood, IL: Dow Jones-Irwin, 1983.

Lieberman, M., L. Lau and M. Williams, "Firm-Level Productivity and Management Influence: A Comparison of US and Japanese Automobile Producers," Research Paper No. 1048, Graduate School of Business, Stanford University. Forthcoming, *Management Science.*

Schonberger, R. J., *Japanese Manufacturing Techniques: Nine Hidden Lessons in Simplicity.* New York: Free Press, 1982.

[6] Note that this may simply reflect regression to the mean.

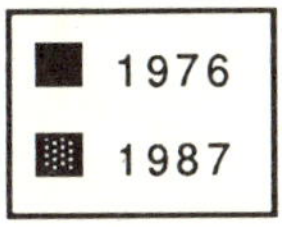

FGI/Sales ratios of automotive parts suppliers

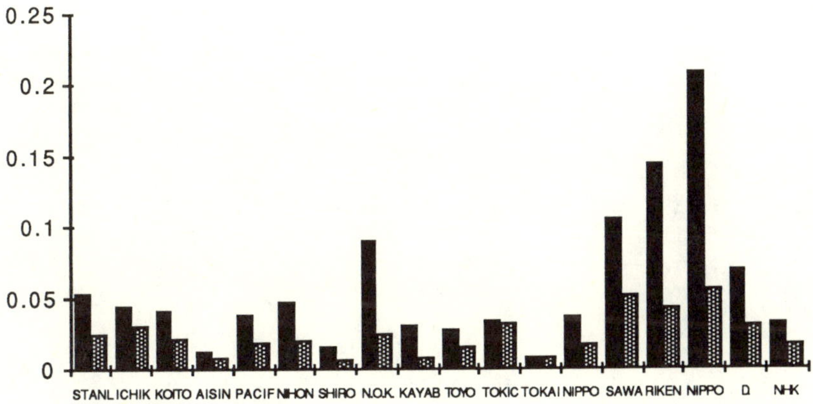

RMI / Materials cost ratios of automotive assemblers

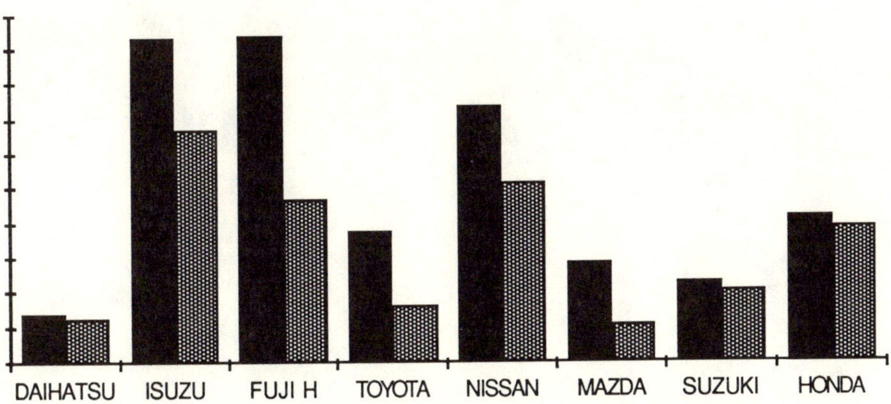

22

INTENDED AND ACHIEVED COMPETITIVE PRIORITIES: MEASURES, FREQUENCIES, AND FINANCIAL IMPACT

Craig H. Wood, Larry P. Ritzman, and Deven Sharma

College of Business, The Ohio State University, Columbus, OH 43210 (614) 292-5233
College of Business, The Ohio State University, Columbus, OH 43210 (614) 292-0258
Booz, Allen and Hamilton, Inc., New York, NY 10178 (212) 697-1900

ABSTRACT

Using data collected from managers in a sample of manufacturing plants, parallel sets of variables representing intended performance and achieved performance are factor analyzed to see if they break out in a similar way. For each competitive priority identified as a factor (latent construct), a frequency table is constructed showing how intended and actual competitive priorities match up in practice. Finally, an overall performance measure is overlaid onto the frequency matrix showing how competitive priorities -- both intended and realized -- affect a firm's bottom-line financial performance.

INTRODUCTION

Manufacturing strategy means patterns or relationships in the stream of manufacturing decisions (Mintzberg 1978). In order to appreciate the importance of these patterns, we must first define good measurements, starting with competitive priorities. We must show that the competitive priorities 'matter' and that they do have an impact on the financial health of the plant.

Competitive priorities are defined by Hayes and Wheelwright (1984) as the dimensions on which a company chooses to compete in the marketplace. The various dimensions of competitive priorities have been the subject of considerable speculation in the operations strategy literature. Wheelwright (1978) identifies four different dimensions: efficiency, flexibility, quality, and dependability. Hayes and Schmenner (1978) recognize differences in flexibility as a competitive priority and list five different dimensions: price, quality, dependability, product flexibility, and volume flexibility. Hayes and Wheelwright (1984) identify a combined list of competitive priorities in their discussion of the subject. Finally, Krajewski and Ritzman (1987) arrive at a more detailed list by differentiating aspects of quality and dependability into two parts each. Their list includes cost, high performance design, fast delivery, consistent quality, on-time delivery, product flexibility, and volume flexibility.

The dimensions suggested as competitive priorities come from a combination of experience with real world operations and knowledge of theoretical concepts in strategic management. Until recently the question of which and how many competitive priorities should be considered was purely a conceptual rationalization. However, the results of several empirical research projects are now available which help to define the type and number of competitive priorities and determine how they can realistically be measured and used to improve the strategic management of manufacturing. Richardson, Taylor, and Gordon (1985), Swamidass (1986), Swamidass and Newell (1987), Sharma (1987), Huete and Roth (1987), Roth (1989), and Wood, Sharma, and Ritzman (1989) all present somewhat different approaches to the problem of measuring the competitive priorities as latent constructs. Latent constructs are attractive for this type of research because they represent the direct and indirect effects on the performance measures (measured variables) which are commonly used in manufacturing.

The core issue of concern here is whether manufacturing managers think in strategic terms when they implement their decisions. By concentrating on the linkage between competitive priorities and performance measures, we can begin to better understand the levers available for planning and controlling manufacturing strategies. This research, although still primarily descriptive in nature, begins to identify the basic rules that govern this linkage process by directly comparing results of past decisions with present intentions of action.

RESEARCH DESIGN AND METHODOLOGY

This research uses survey data collected from manufacturing managers and executives in early 1987 (Sharma 1987). The survey obtained information from 144 different manufacturing plants in the Midwest and Northeast concerning many aspects of the management of their operations. One portion of the questionnaire asked about how the respondents perceived various aspects of their own manufacturing performance compared to their major competitors (see Wood, Sharma, and Ritzman 1989). Measured variables commonly used for product cost, quality, delivery time, and flexibility performance were included. These retrospective measures we have termed "achieved" measures of competitive priorities, because they represent current operating results relative to the rest of the industry.

In contrast, questions at the beginning of the questionnaire asked about what areas manufacturing managers fell are currently most important to them: i.e., in need for immediate improvement or close attention. Respondents were asked to rate the importance of competitive priorities in terms of i) selling their products, ii) as management emphasis, and iii) in evaluating the performance of line managers. These questions are included as measures of "intended" performance: planned competitive priorities which are currently emphasized to develop or maintain competitive advantage.

Table 1 identifies the questions from the survey which are used as measured variables for this analysis. Only those questions which can be matched on both intended and achieved performance have been included in order to prevent the possibility of obtaining latent constructs in the data analysis of one dimension which cannot be matched on the other dimension because the appropriate questions were not included. The variable numbers such as X18 and X151 are used to identify the individual questions in the survey instrument for the data analyses. Based on prior theory, the thirteen intended and eleven achieved measured variables could reflect up to six of the seven potential competitive priorities: none of these variables were expected to load on a volume flexibility construct.

The objective of comparing intended actions with achieved performance requires that data analysis be performed on the two groups of variables separately and then the results combined. Using SAS for all of the statistical results, we first performed correlation analyses and maximum likelihood factor analyses, choosing the best fitting parsimonious solution based on scree plots and the comparison of chi square tests of overall fit and Tucker and Lewis's Reliability coefficient for a range of probable best solutions. For both the intended and achieved performance variables, the four factor solution proved to be best. Oblique (Promax) rotations of each of the two sets of variables for three, four, five, six, and seven factor solutions were run and compared in order to gauge the sensitivity of the various factor loading patterns to the addition and deletion of latent constructs (factors).

The factor scores from the four factor solution were than used as weights to incorporate the measured variables into two-way frequency tables for each of the four latent constructs. This technique allowed us to combine the frequencies of responses with intended performance as the rows and achieved performance as the columns. In this manner, we are able to create a single set of frequencies to use for evaluation of the two different sets of responses.

One goal of our study was to see if we could learn if large discrepancies between achieved and intended performance lead to poor overall performance. To evaluate this, we returned to the data and factor analyzed the set of achieved performance variables expanded to include four measures of overall firm performance: X163 - Market Share, X164 - Sales Growth, X165 - Earnings Growth, and X166 - Profit Margin. Even though an earlier study (Wood et al. 1989) had already established the existence of an overall firm performance latent construct which loaded significantly on these variables, the data was re-analyzed to ensure consistency with our current comparisons of intended and achieved competitive priorities. The factor analysis verified the existence of this factor and the factor scores for the measured variables were used as weights -- in the same way as described above for the intended and achieved competitive priorities -- to develop our measure of financial performance.

Measured Variables Used in the Analysis			
INTENDED		ACHIEVED	
Indicate the importance you attach to each alternative in		Relative to your significant competitors, please indicate your position on the following:	
A. Selling products in your primary product line:			
X16	Low Price	X151	Product Cost
X17	High Product Performance		
X18	High Durability	X152	Product Performance
X19	High Product Reliability	X153	Number of Features
X22	Short Delivery Time	X154	Product Durability
X23	Delivery on Due Date	X155	Product Reliability
X24	Customizing Product to Customer Specification	X156	Quality Consistency
		X157	Quality Perceived by Customer
X26	Number of Features	X158	Delivery Time
		X159	Delivery Dependability
		X161	Ability to Customize
		X162	Product Price
B. Management priorities in manufacturing:			
X27	Production Cost		
X30	Conformance to Design Specifications		
X31	Improve Quality in Manufacturing		
C. As criteria in evaluating your line managers' performance:			
X37	Cost		
X38	On Time Delivery		

Table 1.

The financial performance measure is calculated by summing the products of each financial performance score by the frequency that that score was observed in the intended-achieved competitive priority frequency tables. Financial performance scores range from level 1 for performance significantly worse than the competition to level 5 for performance significantly better than the competition.

Because of the sparseness of data points in the level 1 (meaning Not Important for intended performance and Significantly Lower for achieved performance) and level 2 (Somewhat Important and Somewhat Lower) responses to the measured variables, we collapsed levels 1, 2, and 3 of the frequency tables into a single group of observations for analysis of the results. Thus the 1 - 3 level category.

MEASUREMENT OF COMPETITIVE PRIORITIES

The first step in the analysis is to determine the degree of similarity that exists between intended and achieved competitive priorities. We did this by factor analyzing the two different sets of measured variables shown in Table 1 and then comparing the results to see if different latent constructs appear in the solutions. What we are trying to understand with this line of analysis is when manufacturing managers say that certain competitive priorities are more important than others, do these managers reflect this high level of importance in their operating decisions.

The factors broke out in a very neat and orderly fashion. Table 2 shows the results of the four factor solutions for both the intended and achieved performance data. The common convention of numbering the factors as they appear in the solution when presenting factor analysis results has been followed. This means that before rotation of the axes, the first factor F1 explained the largest amount of variance in the data, the second factor F2 explained the largest amount of the remaining variance after the first factor was removed, and so on. As can be seen in Table 2, both the types of variables loading on each competitive priority cluster (factor) are very similar and the order in which the factors enter the solution is identical for both sets of data. Thus the thesis of similar underlying constructs is supported: manufacturing managers do appear to treat intended and achieved performance in much the same way.

FACTOR LOADINGS - PROMAX ROTATIONS									
Intended Performance	Loadings*				Achieved Performance	Loadings*			
Variable	F1	F2	F3	F4	Variable	F1	F2	F3	F4
X17 - Prod. Perform.	.75				X152 - Prod.Perform.	.57			
X18 - Hi. Durability	.65				X154 - Durability	1.05			
X19 - Reliability	.65				X155 - Reliability	.83			
X26 - No. of Features	.33				X153 - No. of Features	.41			
X31 - Improve Quality	.39								
X22 - Short Del.iv Time		.64			X158 - Delivery Time		.90		
X23 - Del. on Due Date		.84			X159 - Depend. Deliv		.88		
X38 - On-time Delivery		.42							
X16 - Low Price			.46		X162 - Product Price			.42	
X27 - Production Cost			.80		X151 - Product Cost			1.00	
X37 - Cost			.55						
X30 - Conform. to Spec.				1.02	X152 - Product Perform.				.30
					X156 - Qual. Consist.				.59
					X157 - Qual. Perc. Cust.				.78

* Only the factor loadings > .30 are shown.

Table 2

The interpretation of these factors is relatively straight forward and follows closely the results obtained in our earlier studies (Sharma 1987 and Wood et al. 1989). Factor one is the physical or high performance design side of the quality competitive priority; factor two is the delivery or time competitive priority; factor three is the cost/price competitive priority; and factor four is the reputation or quality consistency side of the quality competitive priority. In both sets of data, factors one and four are fairly highly correlated with one another, in fact, in the three factor solution they are combined into a single factor (with different loadings on some variables). One somewhat surprising aspect of this factor solution, is that the flexibility competitive priority does not appear (as it did in our earlier studies). However, this is less surprising than it might at first appear because of the absence of several unmatched variables from the achieved data set which loaded heavily on the flexibility construct in our previous studies.

Several interesting patterns in these solutions should be noted. First the variables reflecting customization of products do not load heavily on any of the factors in either the intended or achieved four factor solutions. This reinforces the parallel-solutions conclusion reached about the factor analyses: customization does not load on either set of factors. Also the absence of customization ties in closely with the missing flexibility factor mentioned above. Measured variables have been left out of the model which we are testing which will not let us "capture" the flexibility construct in our models. This is important for the future redesign of questions in the survey instrument.

Another interesting pattern is the collapse of the two time or delivery dimensions (on time delivery and short delivery times) onto a single latent construct Just the opposite from the quality priority, these hypothesized two separate priorities (Krajewski and Ritzman 1987, pp. 43-49) seem to want to stick together, indicating that you must have both at once: short delivery times alone are not an effective way to compete; you must also have dependability of delivery to have a successful time based strategy. Finally, in the achieved data, because the product performance variable (X152) loads on both the high performance design factor (F1) and the quality consistency factor (F4), we have additional support for the proposition that quality is a construct which has two distinct dimensions to it. For this portion of the study, we successfully demonstrated the similarity of the underlying latent constructs for both the intended and achieved sets of competitive priorities.

PATTERNS IN FREQUENCY TABLES

The factor scores from the four factor solution were generated and used as weights for the measured variables. A set of response frequencies was calculated for each of the factors. Frequency tables were constructed with intended response levels (1 to 5) as the row variables and achieved response levels (1 to 5) as the column variables. The individual cell values are the frequency of response for each achieved level of competitive priority at intended levels of performance. The response frequencies are shown in Tables 3 through 6 below as the top number of the pair in each cell. The numbers in parentheses are the financial performance measures which will be discussed in a later section of the paper. Note that the levels 1, 2, and 3 have been collapsed into a single category. This was necessary due to the sparseness of the data in the 1 and 2 levels. Not surprisingly, almost no one is willing to say that they intend to operate their plant at a very low level on any competitive priority.

One strong pattern in the responses can be seen by looking at the marginal frequencies for all four factors. In all cases the number plants which intend to operate at a high level (4 or 5) relative to the industry average (3).. However, on all factors the average level achieved on the competitive priorities is below the average intended levels. Said in a different way, there clearly is a tendency toward underachieving after setting lofty goals.shown by the patterns in the data.

Finally there are some interesting patterns reflected in the where we see sparse frequency table cells. The diagonal line of cells from upper left to lower right represents congruence of intended and achieved performance. Looking at cells off the diagonal, we can characterize responses above and to the right of the diagonal as those of "overachievers" and responses from below and to the left of the diagonal as those of "underachievers." With this characterization of the frequency tables in mind, we can see that plants do not tend to disperse uniformly over the entire matrix. All factors in our study show significantly higher frequencies of underachievers than of overachievers. In fact, factor 1 - the high performance design priority - has no overachievers at all. The main tendencies exhibited in the frequency data are toward the 'satisficing' of priorities -- having a balance of reasonably high emphasis on several priorities rather than the very highest emphasis on just one -- and underachieving are the main findings in this line of analysis.

PATTERNS IN FINANCIAL PERFORMANCE

Tables 3 through 6 present the financial performance scores in parentheses in the body of the tables and the marginal financial performance by level of intended priority in the column at the far right and by level of achieved priority in the bottom row. The on-diagonal entries have been underlined for enhancing the patterns which exist. These tables show both the position of plants on the response frequency matrix and the financial performance score for the plants in each cell.

Factor 1: High Performance Product Priority

This intended-achieved priority set are shown in Table 3. Being on the diagonal generally produces better financial results on this factor. Financial performance clearly improves as intentions and achievement increase from the 1 - 3 level to the 4 level up to the 5 level. Also, with one exception, it is clear in Table 3 that firms are better off when intended and achieved priority levels match. The exception is in the intended level 5/achieved level 4 (5/4) cell which shows better firm performance (4.00) than plants in the intended level 5/achieved level 5 (5/5) cell. This same pattern shows up in other factors and leads to the conjecture that 5/5 may be too extreme a position to producer superior financial performance: losses in other competitive priorities cannot be compensated for by excellent achievement in one area.

Factor 1: High Performance Design Priority				
	Freq./(Aver. Fin. Perf.) Achieved			Total
Level	1 - 3	4	5	Intended
Intended 1 - 3	9 / (3.33)	0 / (0)	0 / (0)	9 / (3.33)
4	19 / (2.94)	10 / (3.60)	0 / (0)	29 / (3.17)
5	7 / (3.42)	6 / (4.00)	3 / (3.66)	16 / (3.68)
Total Achieved	35 / (3.14)	16 / (3.75)	3 / (3.66)	54 / (3.35)

Table 3.

Factor 2: Delivery or Time Priority				
	Freq./(Aver. Fin. Perf.) Achieved			Total
Level	1 - 3	4	5	Intended
Intended 1 - 3	13 / (3.30)	5 / (3.80)	0 / (0)	18 / (3.44)
4	21 / (3.28)	18 / (3.83)	2 / (4.50)	41 / (3.58)
5	19 / (3.36)	6 / (3.50)	5 / (3.00)	30 ' (3.33)
Total Achieved	53 / (3.32)	29 / (3.75)	7 / (3.42)	89 /(3.47)

Table 4.

Factor 3: Cost Priority				
	Freq./(Aver. Fin. Perf.) Achieved			Total
Level	1 - 3	4	5	Intended
Intended 1 - 3	7 / (3.57)	8 / (3.13)	2 / (3.50)	17 / (3.35)
4	26 / (3.42)	13 / (3.69)	0 / (0)	39 / (3.51)
5	22 (3.32)	10 / (3.60)	0 / (0)	32 / (3.40)
Total Achieved	55 / (3.40)	31 / (3.51)	2 / (3.50)	88 / (3.44)

Table 5.

Factor 4: Quality Consistency Priority				
	Freq./(Aver. Fin. Perf.) Achieved			Total
Level	1 - 3	4	5	Intended
Intended 1 - 3	4 / (2.75)	5 / (3.20)	3 / (3.67)	12 / (3.17)
4	14 / (3.14)	14 / (3.42)	7 / (3.57)	35 / (3.40)
5	14 / (3.64)	23 / (3.69)	5 / (3.20)	42 / (3.61)
Total Achieved	32 / (3.31	42 / (3.54)	15 / (3.46)	89 / (3.44)

Table 6.

The sample size in this factor combination is also quite small compared to the other three factors. This happens because of the relatively large number of variables which were weighted together to obtain the response frequencies for each factor. Observations with missing data on any one of the variables used have been thrown out of the analysis. Several variables in this factor combination had high rates of no response creating missing data points.

Factor 2: Delivery or Time Priority

This second factor pair shown in Table 4 also clearly indicates that at medium levels of emphasis on competitive priorities being on the diagonal contributes to improved overall financial performance. The 1 - 3 and 4 levels show that a firm is clearly better of to have congruent manufacturing priorities than to have either different intended or achieved performance. But again the very best financial performance plants are found of of the 5/5 position. The 5/5 cell shows lower firm performance than either the 5/4 or the 4/5 cell.

Another interesting pattern on this factor is the unexpectedly high firm performance experienced by overachievers. Both the 1 - 3/4 and 4/5 cells show that overachieving on the time priority produces improved performance.

Factor 3: Cost Priority

Table 5 shows the response frequencies and financial performance measure scores for the cost priority pair. Once again the firm is best off financially if it is not in an extreme position. The best situation here is to be on the diagonal, but below the best at this priority. Clearly emphasis on cost to the detriment of the other priorities will not necessarily improve the plant's financial performance.

Factor 4: Quality Consistency Priority

The quality consistency competitive priority pair is shown in Table 6. This table has the most evenly dispersed response frequencies among the four factors. Here, however, there appears to be a situation which says that more emphasis is better -- as long as you are not in the 5/5 position. Once again, medium levels of achievement while striving for excellence helps to improve financial performance. With this factor , quality consistency, we see that it appears to be better to be off the diagonal even at low to medium levels of emphasis.

CONCLUSION

Although the patterns do not seem to exhibit absolute relationships between intended and achieved emphasis on competitive priorities and overall financial performance, several general trends are apparent in the data:

1. Both intended and achieved competitive priorities show the same configuration of latent constructs (factors) underlying the priorities and these factors load on the same measured variables and have the same order of entry into the factor solutions.
2. The higher the intended or achieved competitive priority, the better the firm's performance.
3. The closer the plant's position is to being on the competitive priority-pair diagonal, he better the firm's overall performance (except for quality consistency)..
4. Most firms aim high on each competitive priority, but frequently underachieve their goals.
5. Balance - above average financial performance seems to require above average but not the highest level of emphasis on a wide variety of competitive priorities.

Not only do manufacturing managers feel that it is important to operate at or above some average level of performance on the competitive priorities, in fact they seldom do operate at those levels. For most priorities to be successful in producing competitive advantage, they must be balanced with complementary positions on the other priorities. Future research must focus on defining the most and least successful combinations of competitive priorities and identifying the specific situations in which these combinations should be employed. This study supplies a beginning for developing quantitative measures for these future efforts.

REFERENCES

Hayes, Robert H., and Roger W. Schmenner, "How Should You Organize Manufacturing," Harvard Business Review, Jan.-Feb., 1978, 105-118.

Hayes, Robert H., and Steven C. Wheelwright, Restoring Our Competitive Edge: Competing Through Manufacturing, New York: John Wiley & Sons, 1984.

Huete, Luis, and Aleda Roth, "Linking Manufacturing Capabilities with SBU Strategic Directions," Working Paper, Boston University, School of Management, 1987.

Krajewski, Lee J., and Larry P. Ritzman, Operations Management, Reading, MA: Addison-Wesley Publishing Company, 1987.

Mintzberg, Henry, "Patterns in Strategy Formation," Management Science, 24, 9, May, 1978, 934-948.

Richardson, P.R., A.J. Taylor, and J.R.M. Gordon, "A Strategic Approach to Evaluating Manufacturing Performance," Interfaces, 15, 6, Nov.-Dec., 1985, 15-27.

Roth, Aleda, V., "Linking Manufacturing Strategy and Performance: An Empirical Investigation," Working Paper, Boston University, School of Management, 1989.

Sharma, Deven, Manufacturing Strategy: An Empirical Analysis, Unpublished Ph.D. dissertation, The Ohio State University, 1987.

Swamidass, Paul M., "Manufacturing Strategy: Its Assessment and Practice," Journal of Operations Management, 6, 4, August, 1984, 471-484.

Swamidass, Paul M., and William T. Newell, "Manufacturing Strategy, Environmental Uncertainty and Performance: A Path Analytic Model," Management Science, 33, 4, April, 1987, 509-524.

Wheelwright, Steven C., "Reflecting Corporate Strategy in Manufacturing Decisions," Business Horizons, February 1978, 57-65.

Wheelwright, Steven C., "Manufacturing Strategy: Defining the Missing Link," Strategic Management Journal, 5, 1984, 77-91.

Wood, Craig H., Deven Sharma, and Larry P. Ritzman, "Strategic Information Management: Competitive Priorities and Manufacturing Performance Measures," Proceedings of the Eighth National Conference of the Operations Management Association, Columbus, Ohio, April, 1989.

23

STRATEGIC ENTRY INTO GLOBAL MANUFACTURING INDUSTRIES: EVIDENCE AND NEW THEORETICAL DIMENSIONS

Will Mitchell
Assistant Professor of Corporate Strategy
School of Business Administration
University of Michigan
Ann Arbor, MI 48109-1234
(313) 764-1230

Avi Fiegenbaum
Assistant Professor of Corporate Strategy
School of Business Administration
University of Michigan
Ann Arbor, MI 48109-1234
(313) 764-1349

ABSTRACT

This paper develops a theory of strategic entry groups, entry-site strategic groups, and subsequent mobility across strategic groups within heterogeneous industries. We illustrate the concepts by discussing strategies followed by members of one strategic entry group in the medical diagnostic industry.

INTRODUCTION

Major American firms once met few challenges when they competed in international markets. As industry after industry has become progressively global, however, companies based in the United States have had to change their competitive strategies (Cohen, Teece, Tyson, and Zysman, 1984). Those that have not changed have failed.

In this paper, we will address this issue in two ways. The first part of the paper is empirical. We will describe the globalization of the medical diagnostic imaging industry during the last 30 years, with a focus on the strategies used by Japanese firms to compete with American and European manufacturers. We will then use the empirical case to motivate a theoretical framework for analysis of three specific competitive issues in global manufacturing industries. In the theoretical section we will discuss the origins of entry groups, their entry-site strategic groups, and their later movement across strategic groups.

GLOBALIZATION OF THE DIAGNOSTIC IMAGING INDUSTRY

In this section we describe one global high-technology manufacturing industry and entry strategies used by the entry group of Japanese firms that have competed in its American market. Starting from minor positions in the United States during the 1950s, Japanese manufacturers have expanded to become strong players in the American segment of the international medical diagnostic imaging industry (Foote and Mitchell, 1989). The strategies they have used - how they entered and then how they expanded - underlie their current competitive strength. The description of the strategies will provide an example of the importance of the related theoretical issues.

Innovation and Competition in the Imaging Industry

Diagnostic imaging equipment is used to produce images of physiological structures within the body. Conventional x-ray and electrodiagnostic instruments have been available since early in the century. Since the 1950s, however, nuclear medical, ultrasonic, computed tomographic, magnetic resonance, and digital radiographic instruments have expanded the scope of the industry and challenged industry incumbents (Mitchell, 1988). We first describe imaging equipment development and commercialization trends and then identify competitive challenges mounted by Japanese entrants to the U.S. market.

Each general type of imaging equipment emerged from academic research laboratories throughout the world. Most commercial prototypes were introduced to medical markets by newcomers to the diagnostic imaging industry, often after contact with the academic researchers. Once new types of devices achieved both technical and market success, the major diagnostic imaging industry manufacturers then incorporated the products into their lines.

During the late 1950s, international sales of conventional x-ray and electrodiagnostic equipment totalled about $100 million, with almost half of the sales in the United States. With sales growth of conventional instruments and successive introduction of new methods of producing images, global shipments of imaging systems will exceed $9 billion by 1990, about one third in the United States. Sales of auxiliary equipment and after-sales service will total several times this figure. As shipments have grown, so has the globalization of the industry that produces the devices. Once a series of national industries, the diagnostic imaging industry has now become truly international.

In 1960 the American radiography industry consisted of about 45 firms selling equipment into a $40 million market. The industry was dominated by three American firms, General Electric, the Picker X-Ray Company, and the Westinghouse Electric Company, followed by two European producers, Siemens AG and NV Philips. The five manufacturers shared 70% to 75% of the U.S. market for x-ray equipment and a significant share of the smaller electrodiagnostic instrument market.

In 1990, the radiography industry has become the diagnostic imaging industry, with about 150 firms annually selling $3 billion of x-ray, electrodiagnostic, nuclear, ultrasonic, computed tomography, magnetic resonance, and digital radiographic equipment to American buyers. Among the more than 350 manufacturers that have entered emerging subfields of the industry during the past 30 years, several distinct entry groups can be identified. In addition to the traditional industry leaders, the entry groups include startup firms, unrelated diversifiers, related diversifiers based in Europe and the U.S., and established Japanese firms.

The most successful of the entry groups has been the established leaders. The industry is still dominated by General Electric, Picker (now part of GEC PLC of Britain), Siemens, and Philips. Together, the four firms control about 70% of the American market. Of the 1950s leaders, only Westinghouse, which left the industry in 1971, has dropped out.

Firms in most other entry groups have had little success in the industry. Almost all startups have failed, because of weak financial and organizational capabilities. Most unrelated diversifiers have left the industry after learning that there are few profits for participants lacking technical and sales backgrounds suited to the imaging market. Even the group containing experienced American and European medical and electronic equipment manufacturers that made large investments in imaging research, development, manufacturing, and distribution systems has been signally unsuccessful. Several aerospace and military electronics companies, for example, have manufactured diagnostic imaging equipment, believing that their technical expertise would allow them to succeed; all have failed to become general imaging equipment manufacturers. Half a dozen major drug companies have entered, expecting their medical technical and market experience to give them an edge in the imaging industry; none is now an imaging market leader and most have left the industry.

Of all the entry groups to the U.S. industry, only the group consisting of experienced Japanese manufacturers has produced broadly-based imaging equipment manufacturers that are likely to enjoy increasing success in the industry. The Toshiba Corporation, Hitachi Ltd., and the Shimadzu Corporation are beginning to approach the status of the historical market leaders and each has the potential to expand from its existing base. Only the interaction between nationality of firm ownership and technical and financial capability can explain the continued success. The Japanese entry group had a successful combination of firm and national-level capabilities.

Five Japanese Entry and Mobility Strategies

Most Japanese entrants to the global imaging industry have been large, diversified firms, often part of networks of companies possessing broad bases of financial, technical, manufacturing, R&D, and distribution strength. Thus, they have had the resources needed to build long-term positions in the industry. The firms have used those resources to enter and expand within the U.S. market through a sequential series of five strategies. The strategies, which include the firms' entry-site strategic groups and then movement across strategic groups, are listed in Table 1.

Table 1

Key aspects of Japanese firm entry-site and expansion strategies

	Key aspects of strategy	Time first used *
1.	Sell components to American and European manufacturers	1950s
2.	Sell systems to American and European manufacturers	1960s
3.	Sell Japanese-branded systems	
	a. Indirect distribution	late 1960s
	b. Direct distribution	late 1970s
4.	Japan-based manufacturing joint ventures with American and European manufacturers	1980s
5.	U.S.-based assembly plants; most value-added in Japan	late 1980s

* Each strategy continues, but supplemented by later strategies.

The first strategy is the supply of components that are incorporated into imaging systems manufactured by European and American firms. Japanese manufacturers have acted in this role since the 1950s. The practice is found in all technical subfields of the industry, from the oldest conventional x-ray products to the newest positron emission tomographic nuclear medical scanners.

The second strategy is the supply of systems that are sold with European and American firms' labels. Japanese manufacturers have played this role since the 1960s. The activity is most common in the older subfields such as x-ray, electrodiagnostic, nuclear, ultrasound, and computed tomography. It is also common in niches within those subfields. In x-ray mammography, for example, Acoma X-ray Industry Co. Ltd. supplies about 25% of the systems sold in the U.S. Japanese manufacturers are also moving from the established subfields into the newer product areas, as Japanese firms now supply magnetic resonance and digital radiography systems sold by American and European firms.

The third Japanese manufacturing mobility strategy is the sale of Japanese-branded products through indirect and direct distribution networks. Indirect distribution through sales representatives was the norm until the mid-1970s. Even now, many Japanese firms sell products through distributor networks in the U.S.; in 1988, for example, Acoma began to contract with American dealers to sell its mammography units. Indirect distribution is not an effective way of selling complex equipment, however, and the market share of Japanese-firm branded products often has been small. In the less-complex electrodiagnostic instrument area, though, products manufactured by Japanese firms such as the Nihon Kohden Corporation and NEC-San Ei Instruments Ltd. dominate the market.

Since the late 1970s, starting with Toshiba in 1976, several Japanese medical equipment manufacturers have established direct sales systems. This is the strategy of at least the three strongest Japanese competitors, Toshiba, Hitachi and Shimadzu. Although the direct sales presence of these firms is not yet large, market shares will increase as the sales and service systems become better established.

The fourth Japanese manufacturing strategy is the establishment of joint ventures in Japan between Japanese and American or European companies. Japanese firms enter into the joint ventures in order to gain access to technical, market, and regulatory knowledge held by the American and European firms. The Western manufacturers, in turn, enter the ventures to acquire high quality production, technical advances, and lower cost manufacturing capability. Although a few American medical equipment companies have long had manufacturing joint ventures in Japan, the number of joint ventures recently has increased.

Most major European and American imaging manufacturers now participate in Japan-based joint ventures with Japanese firms. Yokogawa Medical Systems (YMS), a 1982 venture of General Electric and the Yokogawa Hokushin Electric Corp., is the strongest of these. YMS sits with Toshiba and Hitachi atop the Japanese imaging market and, through GE, is becoming an important supplier to the U.S. market. YMS now manufacturers most of GE's ultrasound product line, much of its computed tomography line, and in the last three years has also begun to manufacture magnetic resonance imaging systems. Picker-Toray-Fuji, a joint venture of Picker, Toray Industries, Inc., and the Fuji Electric Co., is now responsible for most of Picker's ultrasound and nuclear production and is also developing magnetic resonance imaging systems. Siemens, too, has recently formed a joint venture with the Asahi Optical Co. Ltd.

And a fifth stage of the Japanese manufacturing strategy is the establishment of manufacturing plants in Europe and the United States. Toshiba, for example, has established a California assembly plant for some of the imaging instruments that it sells in the American market. Thus far, however, most of the advanced manufacturing and consequent value-added takes place in Japan.

If the Japanese firm had attempted to jump directly to the senior strategies, they probably would have failed. The firms would not have understood the U.S. market (Toshiba's first ultrasonic scanners, for instance, were too small for many American patients) and potential users would not have understood the firm. But by taking the sequential road of entry and expansion across strategic groups, firms such as Toshiba and Hitachi gradually have moved into strong positions. At the same time, because the earlier strategic steps tend to be less expensive than the later stages, Japanese firms that have not been as broadly successful as Toshiba and Hitachi have limited their entry costs. The five stage sequence, therefore, has been both a ladder and a toe. For successful firms, it has provided a ladder to the top. And for less successful manufacturers, it has provided a low-cost way of testing the U.S. waters.

The previous discussion has focused on a key entry group in the diagnostic imaging industry, the entry group of established Japanese manufacturers that have expanded into American markets. Many of these firms have become important competitors in the industry. However, we do not have a systematic way of analyzing the strategies used by such companies to gain competitive advantage.

In the next section, we move from induction to deduction. We review important ideas from the economics and management theory entry literature. Three components of our empirical discussion - entry groups, entry site, and later movement - then form the basis of new theoretical dimensions of entry strategy. Our perspective extends the existing literature, helping to identify entry and expansion strategies and understand why such strategies may be successful.

REVIEW: ENTRY STUDIES AND INDUSTRY STRUCTURE

Entry into new industries has received attention from many researchers in economics and management. Analysts have examined entry strategies such as entry mode (Caves and Mehra, 1986; Kogut and Singh, 1987) and entry timing (Lieberman and Montgomery, 1988; Mitchell, 1989b). The researchers have argued that the strategies will be affected by country characteristics (Ajami and Ricks, 1981), entrant characteristics (Montgomery and Hariharan, 1987), exogenous industrial entry barriers (Bain, 1956), and entry prevention by industry incumbents (Gaskin, 1970). Similarly, subsequent performance will be influenced by entry strategies (Mitchell, 1989a).

The internal structure of the industry into which entry occurs will also affect entry strategies. Hunt (1972) coined the term "strategic group" to define a set of firms within an industry that follow similar strategies. The concept has been used to analyze the structure of complex industries (McGee and Thomas, 1986; Schendel and Patton, 1978; Cool and Schendel 1988; Mascarina, 1989), define competitors (Porter 1980; Fiegenbaum and Thomas, 1989), and examine the range of competitive positions available within an industry (Cool and Schendel, 1987, 1988). Fiegenbaum and Thomas (1989) also have developed a theory which argues that a strategic group acts as a reference group for group members.

Caves and Porter (1977) generalized the idea of entry barriers into the concept of mobility barriers, which act as barriers within an industry. Porter (1979) then proposed that mobility barriers will allow certain strategic groups to maintain persistent performance advantages over other strategic groups. Empirical studies have found partial support for this theory (Newman, 1973, 1978; Oster, 1982; Cool and Schendel, 1987, 1988; Fiegenbaum and Thomas, forthcoming).

In this section, we have noted that entry strategies will be affected by characteristics of individual firms. We also have argued that post-entry behavior and performance will be influenced by mobility barriers between strategic groups. The next section will describe how the concepts of strategic groups and mobility barriers can be combined with entry analysis to understand aspects of the strategic entry and subsequent movement within heterogeneous industries.

THEORY: STRATEGIC ENTRY AND SUBSEQUENT MOBILITY

Origins of Entrants: Strategic Entry Groups

Where do new entrants come from? Are they arbitrarily selected from a pool of all possible firms? We argue, instead, that there are strategic profiles of entrant characteristics, which we will call strategic entry groups. And just as membership in a strategic group of current competitors may explain performance outcomes, so may membership in a strategic entry group explain entry outcomes.

Porter (1980) noted that potential entrants are an important industrial force. His five forces model, however, does not help to identify strategic entry groups. But if we rearrange the five forces model into our cross-currents model of industrial fortune, shown in Figure 1, it now becomes possible to classify potential entrants.

The cross-currents model is derived from Schumpeter's (1934) observation that technological change may occur through innovation in components, production processes, products, uses, or organization. Like Porter's (1980) five forces model, we argue that industries are affected by upstream, midstream and downstream forces, that is, by supplier, existing producer, and customer attributes. We also identify a product cross-stream, which includes producers of substitute products but also adds manufacturers of complementary goods and firms with experience in foreign markets. Moreover, we add a social cross-stream category, which includes the effect of laws, customs, and social norms on industrial health. The model encompasses the five forces framework, therefore, but augments it with key aspects of industrial fortune.

Unlike the five forces model, we do not treat potential entrants as a distinct category of industry-level analysis. Instead, we recognize that entrants usually emerge from the other key categories of industrial definition. New entrants sometimes are suppliers or customers for existing products or components; they may be producers of items that will substitute for or complement existing goods; and they sometimes emerge from the social cross-stream when, for example, a government nationalizes an existing member of an industry or forms a new competitor. We expect, therefore, that by analyzing the cross-currents that are pummelling an industry, we will be able to identify many of the potential entrants to it.

In addition to determining where an entrant comes from, by classifying entrants into strategic entry groups, it is important to examine where an entrant will first land in an industry and where it will later move. In the next section, we outline the concepts of entry-site strategic groups and sequential mobility into subsequent strategic groups.

Entry-Site Strategic Groups and Sequential Mobility

Although the concept of strategic groups and mobility barriers was first used to explain intra-industry performance differences, Caves and Porter (1977) suggested that they also can be used to understand competitive dynamics. A strategy of entry and sequential movement may lower the total cost of overcoming mobility barriers into the strategic group that is the ultimate target of the potential entrant. Costs can be lowered by accumulating knowledge and brand identification in the industry through entry into the initial group. Managerial talent can be developed in a more measured way in this fashion. In addition, the reaction of existing firms to entry may be tempered by such a sequential strategy.

Figure 1

Cross-Currents of Industrial Fortune

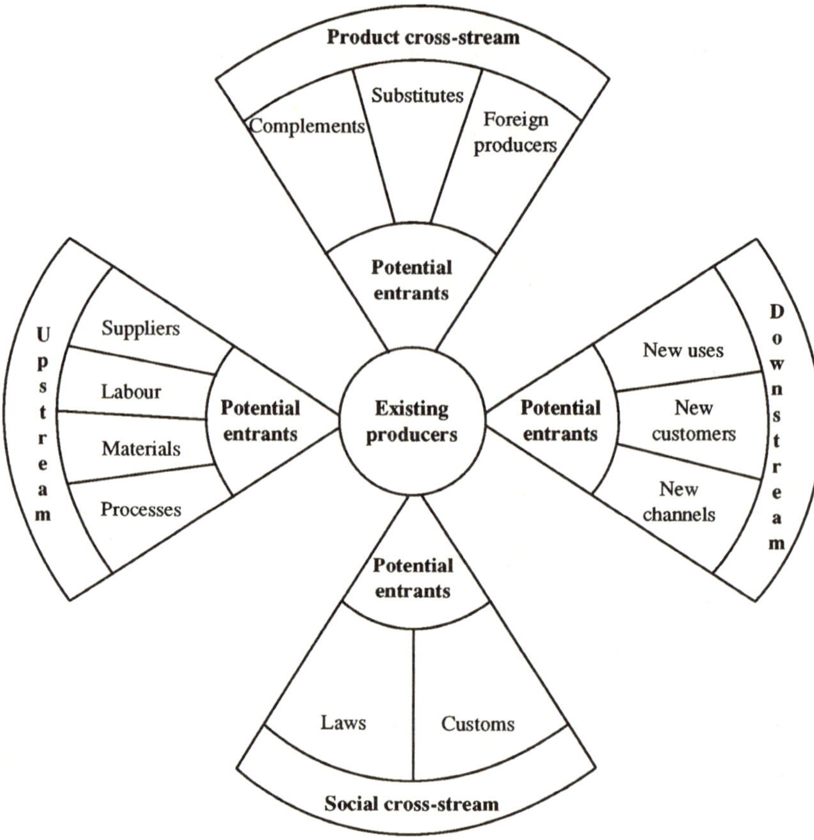

A sequential strategy often lowers the risks of entry because the firm can better control it. If it fails in its initial entry, for example, the firm is spared the cost of going further. In addition, a firm can choose to take its first step into a strategic group in which overcoming mobility barriers requires relatively reversible investments.

Although a firm may benefit by following a sequential strategy, several organizational and economic theories suggest that firm movement within an industry will not be common. Oligopoly theory (reviewed in Scherer, 1980 and Waterson, 1984) argues that oligopolistic firms will behave so as to jointly maximize profits; thus, they may tend to remain within traditional strategic groups rather than expand into new groups. Firms also may remain in the same strategic group over time because of the existence of strategic barriers to change, such as cost (Caves and Porter, 1977) and difficulty of imitation (Rumelt, 1981). Bureaucratic and structural inertia (Crozier, 1964; Hannan and Freeman, 1977) also may inhibit adaptation and create resistance to change.

Thus, there are two broad competing theories about the best entry and movement strategy within an industry. On one hand, some analysts (e.g. Caves and Porter, 1977) have argued that firms should alter their position within the industry by entering different strategic groups over time.

On the other hand, oligopoly, mobility barrier, and organizational theorists argue that there are market and firm-specific impediments to movement.

In order to reconcile these competing theories, we argue that it is necessary to examine strategic entry groups and the capabilities of the firm's within them. Some groups will be able to overcome the mobility barriers of the industry. Other combinations of cross-currents characteristics, however, will not allow a firm to move from an inferior strategic group to a superior one.

CONCLUSION

Firms now competing in global industries face strong challenges from new entrants and expanding competitors, challenges that will only become stronger in the future. We have shown, through empirical evidence from the medical diagnostic imaging industry, how the competitive challenge from Japanese manufacturers has evolved. We argue that the theoretic concepts of entry group, entry-site strategic group, and mobility across strategic groups help us understand the entry strategies followed by Japanese companies.

REFERENCES

Ajami, R.A., & D.A. Ricks, "Motives of non-American firms investing in the United States." Journal of International Business Studies, 12, 1981, pp. 25-34.

Bain, J., Barriers to New Competition, Harvard University Press, Cambridge, MA, 1956.

Caves, R.E., & M.E. Porter, "From entry barriers to mobility barriers." Quarterly Journal of Economics, 91, May 1977, pp. 241-261.

Caves, R.E., & S.K. Mehra, "Entry of foreign multinationals into U.S. manufacturing industries." In M. Porter (ed.), Competition in Global Industries, Harvard Business School, 1986, pp. 449-483.

Cohen, S., D.J. Teece, L. Tyson, & J. Zysman, Competitiveness. Volume III of Global Competition: The New Reality. Working paper of the President's Commission on Industrial Competitiveness (November 8), 1984.

Cool, K., & D. Schendel, "Performance differences among strategic groups." Strategic Management Journal, 9, 1988, pp. 207-223.

Cool, K., & D. Schendel, "Strategic group formation and performance: The case of the U.S. pharmaceutical industry, 1963-1982." Management Science, 33 (9), 1987, pp. 1102-1124.

Crozier, M., The Bureaucratic Phenomena, University of Chicago Press, Chicago, 1964.

Fiegenbaum, A. & H. Thomas, "Strategic groups as reference groups: Theory, modeling, and empirical examination." University of Michigan Working Paper, 1989.

Fiegenbaum, A., & H. Thomas, "Strategic groups and performance: The U.S. insurance industry, 1970-1984." Strategic Management Journal, forthcoming.

Foote, S.B., & W. Mitchell, "Selling American medical equipment in Japan." California Management Review, 31 (4), 1989, pp. 146-161.

Gaskin, W.D. "Dynamic limit pricing: Optimal pricing under threat of entry." Journal of Economic Theory, 3, 1970, pp. 306-322.

Hannan, M.T., and J.H. Freeman, "The population ecology or organizations." American Journal of Sociology, 82, March 1977, pp. 929-964.

Hunt, M.S., "Competition in the major home appliance industry, 1960-1970." Unpublished Ph.D. Dissertation, Harvard University, 1972.

Kogut, B., & H. Singh, "Entering the United States by acquisition or joint venture: Country patterns and cultural characteristics." Journal of International Business Studies, 1987.

Lieberman, M.B., & D.B. Montgomery, "First-mover advantages." Strategic Management Journal, 9, 1988, pp. 41-58.

Mascarina, B. "Strategic group dynamics." Academy of Management Journal, 32 (2), 1989, pp. 333-352.

McGee, J., & H. Thomas, "Strategic groups: A useful linkage between industry structure and strategic management." Strategic Management Journal, 7, 1986, pp. 141-160.

Mitchell, W., "Dual clocks: Entry order influences on performance in industrial markets when specialized assets are accounted for." University of Michigan, School of Business Administration, 1989a.

Mitchell, W., "Whether and when? Probability and timing of incumbents' entry into emerging industrial subfields." Administrative Science Quarterly, 34 (2), June 1989b, pp. 208-230.

Mitchell, W., "The diagnostic imaging industry." Unpublished working paper, The University of Michigan, School of Business Administration, 1988.

Montgomery, C.A., & S. Hariharan. "Diversified entry by established firms." Northwestern University Working Paper, 1987.

Newman, H.H., "Strategic groups and the structure performance relationships: A study with respect to the chemical process industries." Unpublished Ph.D. Dissertation, Harvard University, 1973.

Newman, H.H., "Strategic groups and the structure performance relationship." Review of Economics and Statistics, 60, 1978, pp. 376-383.

Oster, S., "Intra-industry structure and the ease of strategic change." Review of Economics and Statistics, 64, 1982, pp. 376-383.

Porter, M.E., Competitive Strategy. New York: The Free Press, 1980.

Porter, M.E., "The structure within industries and companies' performance." Review of Economics and Statistics, 61, 1979, pp. 214-219.

Rumelt, R.P. "Toward a strategic theory of the firm." Paper prepared for a conference on non-traditional approaches to policy research, Graduate School of Business, University of Southern California, 1981.

Schendel, D.E., & R.G. Patton, "A simultaneous equation model of corporate strategy." Management Science, 24, 1978, pp. 1611-1621.

Scherer, F.M., Industrial market structure and economic performance, Rand McNally, Chicago, 1980.

Schumpeter, J.A., The Theory of Economic Development, Harvard University Press, Cambridge, MA, 1934.

Waterson, M., Economic Theory of the Industry, Cambridge University Press, Cambridge, England, 1984.

24

THE PRODUCTIVITY IMPROVEMENT PROCESS: (PIP) INTERNATIONAL PAPER'S INITIATIVES TO TRANSFORM A MANUFACTURING ORGANIZATION

Gordon A. Enk, Nelson M. Fraiman, Dennis J. Colley and John T. Dillon

International Paper
Long Meadow Road
Tuxedo, NY 10987
(914) 345-5000

I. BACKGROUND AND PURPOSE OF THE PRODUCTIVITY IMPROVEMENT PROCESS

In the fall of 1987 the senior management of International Paper Company initiated a project referred to as the Productivity Improvement Process (PIP). The purpose was to transform an existing operating facility (the Cincinnati folding carton plant), into a leader in its industry. The original objective of the project was "to make the Cincinnati plant (CP) the "best" Folding Carton organization in the industry and to have it serve as a model for other International Paper plants." It was understood at the time that CP was a mediocre performer by most accepted measures (productivity, profitability, quality, safety, receptivity to innovation). At the same time the management and work force had shown consistent improvement and a "can do" attitude towards change. Senior Management turned to the Corporate Technology Organization and chartered the Operations Analysis/Industrial Engineering (OA/IE) group to play a key role in organizing and energizing PIP. The OA/IE group would also serve as the conduit to the entire corporate Technology organization and its staff of more than 500 scientists and engineers. The OA/IE group would have a partnership role with the plant and business group management team in providing leadership and management for the Project. From the outset the project was seen as a prototype effort that would serve as a model to be refined and applied selectively to many of the other 64 converting facilities owned by the company.

The Cincinnati Folding Carton plant employs approximately 300 unionized workers. It had approximately $40 million in sales in 1988 based on a product mix emphasizing frozen food, personal hygiene, and dry food packaging. It has four printing presses and operates three shifts, five days a week. CP is a medium sized facility by International Paper standards based on sales and complexity.

II. INTERVENTION APPROACH - ORGANIZING AND IMPLEMENTING INNOVATION

In April 1988, PIP was launched with a communication program to all employees at CP. The project was described as a partnership between:

1. Hourly Employees (Unionized)
2. Plant and Business Group Management
3. Corporate Technology

Its activities started in June 1988 with a two-day planning workshop involving more than 50 individuals and utilizing the "facilitated" Task Team approach that had become part of the corporate culture in 1984 through the company's Quality Improvement Process (QIP). The workshop participants embraced a "systems approach" for PIP and identified 18 analysis and planning tasks that would be simultaneously addressed by Task Teams. These included:

1. Competitive Environment
2. Product Design and Development
3. Converting Technologies
4. Converting Materials
5. Workflow and Methods
6. Standards and Estimating
7. Production Planning
8. Process Quality Control
9. Inventory Management
10. Purchasing
11. Maintenance
12. Cost Accounting
13. Production Monitoring
14. Statistical Quality Control
15. Waste Management

16. Layout, Warehousing, Building Facilities
17. Materials Handling, Environment, Energy
18. Human Resources

Each Task Team was to be composed of (5-10) individuals from labor, management and Technology. Each Team developed a clear Charter Statement (statement of purpose and focus), and schedule for meetings and results. Task Team meetings were initiated in July. In keeping with the QIP approach, a PIP Lead Team was established and included the:

1. Executive Vice President - Packaging Business

2. Senior Vice President - Technology

3. General Manager - Folding Carton

4. Manager of Manufacturing - Folding Carton

5. Director Product Development and New Business Opportunities

6. Manager of Operations Analysis & Engineering. The original PIP mandates assumed a 4-phase approach to be completed within 24 months.

1.	Analysis and Evaluation	5 months
2.	Preliminary Implementation	7 months
3.	Total Plant Implementation	12 months
4.	Dissemination to Other Converting Facilities	ongoing

Each Task Team was directed to:

- systematically analyze their topic (current status, problems and opportunities);

- assess the potential of new and emerging technologies;

- develop a set of recommendations to move CP to a position of leadership in the industry;

- propose a timetable and strategy to implement the new technology and/or necessary organization changes.

Each Task Team included an OA/IE group staff member who was dedicated to PIP. This staff member worked with a number of related teams and was responsible for ensuring that the Team was well organized, that momentum was maintained, and that sound analysis and information was at the foundation of each project component.

The authors were deeply involved in the planning and conduct of PIP. They were guided by a model of Technology Transfer Intervention (TTI) which has 9 components.

1. STRATEGY -- The TTI must be included in a clearly articulated manufac-turing and business strategy that is shared and understood by all employees.

2. OBJECTIVES -- The TTI must have clearly stated objectives that are quantitative in nature (e.g., profit improvements or productivity gains, quality improvements, or markets share gains).

3. EDUCATION -- The TTI must include the education of all employees on all new technologies and all new approaches to employee involvement and decision making. They must understand the application and principles of the technologies and the organizational culture changes that will take place. Education is basically intellectual and conceptual in nature.

4. TRAINING -- The TTI must include training of all employees on the new technologies so that they have "hands-on" experience before their implementation in a production setting. Training is basically experiential in nature and deals with the application of the education in step 3.

5. INVOLVEMENT AND PARTICIPATORY DECISION MAKING -- Employees must be involved in the data gathering, analysis, planning, decision making and implementation and must have a sense of ownership of the TTI -- it must be done with them, not to them.

6. MANAGEMENT REINFORCEMENT -- Senior and middle managers must provide continuous management reinforcement and communicate a sense of importance and commitment to

the TTI. This is best done through continuous monitoring of the implementation and utilization schedule and by asking managers, supervisors and hourly employees probing questions regarding the technology, its application, and usefulness. It may also include a variety of forms of recognition which are tangible and intangible in nature.

7. BENEFITS SHARING -- Management must be willing to share the benefits of the successful implementation of the TTI with all employees. This can be done through bonuses, profit sharing or gain sharing. The benefits can be on an individual, team or plant-wide basis.

8. EVALUATION -- There must be an ongoing effort to systematically evaluate the progress of the TTI on a monthly and quarterly basis -- the results of which must be shared with all involved parties. The use of quantitative evaluation techniques is of great benefit and should focus on pragmatic factors such as efficiency, productivity, quality, safety, profitability and level of involvement.

9. REDIRECTION AND LEARNING -- There must be a willingness to learn from the process itself and to continuously re-evaluate and re-direct the efforts of the project. Flexibility and responsiveness to emerging opportunities is critical.

This model represents a framework to examine the progress of PIP at CP.

III. RESULTS TO DATE

This paper assesses the results of PIP after its first complete year. Since the project was a first of its kind in the company, the initial 12 months included a significant amount of learning and demanded flexibility and a willingness to learn on the part of all participants.

Table I documents the quantitative results in terms of productivity increases and dollars of cost saving from the initial implementation steps included in PIP.

TABLE I

CINCINNATI PLANT
SUMMARY OF PIP SAVINGS ($000)

Project	1989	1990 Estimate	Capital Investment
Product Design	75	165	125
Converting Materials	22	47	-
Maintenance	37	210	30
Converting Technologies	25	660	1,200
Quality Control	-	30	70
Production Planning	25	260	250
Inventory/TJIT	25	250	-
TOTAL	209	1,610	1,665

These results were a surprise to the project planners who originally assumed that it would take longer than one year to begin to impact the profitability of the plant. The direct costs associated with the first two years of PIP will be $750K. There are, obviously, also a substantial amount of indirect and opportunity costs associated with the project. However, the Return on Investment from the first two years is around 75% and payback is less than two years.

In addition, there was significant qualitative gains as a result of the PIP effort. These primarily involved changes in the attitude and approach taken by management employees to the concept of change. After some initial resistance, managers understood that change was a reality, that it meant the difference between the facilities success or failure, and that new technologies would ultimately improve the work environment.

Unfortunately, coincidental to PIP's involvement in CP, the hourly workforce underwent a prolonged period of labor negotiations which resulted in their refusing any participation. Such participation is obviously critical to achieving such a program's full potential.

25

RADICAL MANUFACTURING STRATEGIES FOR FAILING FIRMS

Paul M. Swamidass

College of Business and Public Administration
University of Missouri
Columbia, MO 65211
(314) 882-2060/882-7374

ABSTRACT

In many ways, failing manufacturers are very different from going manufacturers. Manufacturing strategy process and content in failing firms are dictated by the unique crisis circumstances that prevail in these firms. The resulting manufacturing strategy necessary to turnaround the firm is called turnaround or radical manufacturing strategy (RMS). This paper proposes a framework to organize preliminary ideas and stimulate research and practitioner interest on the issues surrounding RMS. This paper is a report on an ongoing study on business failures.

INTRODUCTION

The premise of this study is that failing manufacturing firms, or manufacturers in serious distress situation represent conditions that call for radical, or turnaround manufacturing strategies. The crisis may be real or anticipated by the company. If the distress conditions are anticipated rather than real, the firm may have more time and resources to deal with the impending crisis. Developing manufacturing strategy for anticipated crises is relatively easier than developing manufacturing strategies for companies in the midst of a crisis. Recent reviews of manufacturing strategy literature show that the literature on manufacturing strategy has not formally addressed the issue of radical manufacturing strategy (Adam and Swamidass, 1989; Anderson, Cleveland, and Schroeder, 1989; Swamidass, 1989).

The proposition of this study is that radical manufacturing strategy calls for quantum changes in the organization at a bad time when it is constrained by liquidity, capital, and time while faced with the risk of failure. These circumstances associated with RMS distinguishes it greatly from the circumstances that exist in going firms. Thus, RMS deserves a special attention from researchers and practitioners.

This paper is organized as follows: first, the causes of failure are discussed, secondly, the question of how failure may be identified or predicted is addressed, and finally, radical manufacturing strategy options for failing firms is presented.

CAUSES OF DISTRESS

The cause of the crisis should dictate the strategy to revive the firm to health. Thus, the most important determinant of manufacturing strategy for firms in crisis may frequently be the cause, or the causes for the crisis. While there is no systematic documentation of the causes of manufacturing crisis, there is some documentation of the causes of business failure. For example, Dun and Bradstreet Corporation (D & B) publishes an annual report called the Business Failure Record, which records all U.S. business failures in the U.S. by cause of failure. The causes of failure associated with manufacturing firms are categorized by D & B as, General, Disaster, Neglect, Fraud, Economic, Sales, Expenses, Customer, Assets, and Capital. Some of the causes of failure used in the categorization of failures by Dun and Bradstreet are too broad and others do not refer to problems with the manufacturing end of the business. Thus, in the interest of RMS research and practice, there is need for research to identify and categorize manufacturing related causes of failure.

The causes of distress in manufacturing businesses (Abernathy, Clark, and Kantrow, 1883; Harrigan, 1988; Porter, 1980; Swamidass, 1989) can be attributed to: (1) change in product design and technology, (2) change in process technology, (3) changes in the market, (4) entry of new competitors, (4) pre-emptive major changes introduced by competitors, (5) product's life cycle reaching the decline stage, (6) excess capacity in the industry, (7) high inventories, and (8) high cost of goods sold. The last two items have been associated with manufacturing business failures (Swamidass, 1989).

HOW TO IDENTIFY FAILING MANUFACTURERS

Failure prediction is discussed in this paper because manufacturing strategy options may become more and more restricted as the firm gets closer to the time of impending failure. Failure prediction could give valuable time for action to practitioners.

The disciplines of finance (Altman, 1968; Blum, 1974), accounting (Deakin, 1972; El Hennaway and Morris, 1983) and banking (Moses and Liao, 1987) study business failures formally. The common theme that spans failure studies in these disciplines is the prediction of business failure using accounting/financial ratios; although the practical value of such research in different disciplines do vary. For example, in banking, the object of failure prediction is to help lenders assess credit worthiness of commercial borrowers, and to assess the risk associated with loans in bank portfolios. In accounting, failure prediction using accounting data helps accounting firms to alert their clients to impending danger.

Business failure prediction research is rooted on the assumption that "impending failure might be symptomized by accounting data" (Blum, 1974, p. 3). The work by Altman (1968) was a turning point in the study of business failure, particularly its prediction. Altman used multiple discriminant analysis (MDA) to find a linear function that could discriminate effectively between going and failed firms. Although, subsequent researchers have used MDA as well as probit, logit and other models to distinguish between failing and going firms, two decades after Altman's original work, the most prevalent approach to business failure prediction is still based on multiple discriminant analysis of selected financial ratios because MDA based predictions have proved to be extremely accurate. For example, in the year before bankruptcy, Altman's benchmark model (1968) was able to predict bankruptcies with a 95% accuracy.

Altman's 1968 failure prediction model, which now serves as a benchmark for failure studies, was based on an index derived from a linear combination of the five following accounting/financial ratios:

1. Working capital/total assets.
2. Retained earnings/total assets
3. Earnings before interest and taxes/total assets.
4. Market value of equity/book value of total debt.
5. sales/total assets.

The above model does not use a manufacturing related ratio. However, a recently completed study (Swamidass, 1989) showed that a multiple discriminant model that included inventory levels, and cost of goods sold was comparable to Altman's benchmark model in predicting manufacturing business failures in selected groups of industries; the five ratios that specified the recently developed discriminant model were:

1. Working capital/total assets.
2. Retained earnings/total assets.
3. Market of equity/book value of total debts.
4. Total inventory/sales.
5. Cost of goods sold/sales.

The fact that (total inventory)/(sales) and (cost of goods sold)/(sales) are associated with manufacturing business failures should be taken into account in developing turnaround manufacturing strategies for manufacturers in distress.

There are effective tools available to practitioners to identify distress in manufacturing firms. Once identified as a distressed firm, manufacturers are faced with the task of dealing with it. The next section briefly considers the options.

MANUFACTURING STRATEGY FOR FAILING FIRMS

Business strategy literature on failing businesses stresses the need for regeneration strategies. Before manufacturing strategy options for failing firms could be discussed, the following must be noted. First, failing firms are often limited by liquidity, capital and time, which in turn limit their strategic options. For example, in the late seventies, Chrysler corporation could not finance its regeneration without congressional backing for its borrowing. Secondly, in some cases, an early exit from the business, rather than regeneration, may be the best decision (Harrigan, 1988).

Corporate distress could be due to uncompetitiveness, endgame conditions (Harrigan, 1988), or crisis brought on by poor internal strategy or uncontrollable external threats. Table 1 summarizes three radical manufacturing strategy options available for manufacturers faced with failure and their impact on known manufacturing strategy variables. The manufacturing strategy variables identified in the table represent variables recognized in literature (Adam and Swamidass, 1989; Swamidass, 1989).

Table 1 implies that there are actually three major individual options or their combinations available to manufacturers faced with a failure. The table does not address financial, marketing and other functional strategy options. The table suggests

Table 1: Options For Turnaround Manufacturing Strategies

	Manufacturing Strategy Dimensions Affected							
Strategy Options	Cost	Quality	Flexibility	Technology of Process	Delivery	Service	Inventory	Focus
1. Plant modernization	x	x	x	x				x
2. Product redesign	x	x	x	x				x
3. Flow	x	x	x		x	x	x	

that the nature of the problem facing the manufacturers must dictate the strategy option or options to adopt. For example, if delivery, service and inventory are major problems, then a strategy that improves the flow must be adopted.

Plant Modernization option is a capital intensive option, and to some extent, Product Redesign option is also capital intensive. If capital in the form of debt or equity is not a viable option, the manufacturer is faced with prompt liquidation or merger or other form of plant disposal.

Table 1 is a framework to identify the issues surrounding turnaround manufacturing strategies. This paper introduces the readers to the concept of radical or turnaround manufacturing strategies and encourages them to recognize the fact that RMS should be discussed and developed. Further, this paper is a reminder that manufacturing strategies in healthy manufacturing firms have the benefit of time, liquidity and capital to work with while developing manufacturing strategies; the same is not true for failing firms.

Table 1 implies that RMS options tend to provide improvement of a subset of manufacturing strategy variables. For example, according to the table, Plant Modernization may impact at least five different manufacturing strategy variables. Consequently, the changes in the manufacturing organization are very pervasive and radical.

SIGNIFICANCE OF THIS STUDY

During the last fifteen years, or so, U.S. manufacturing industries lost their market shares or their entire markets to imports. This means that many U.S. manufacturers simply failed. Examples of failing manufacturers, who eventually survived are: Chrysler, Facit, and Harley Davidson, who made radical strategic decisions for the sake of survival.

But, the entire literature on manufacturing strategy seems to address only the strategy of apparently healthy firms. But, firms faced with a crisis may call for a manufacturing strategy that accounts for at least three major realities:

1. The nature and extent of the crisis in the firm.
2. The reason(s) for the crisis.
3. The limitations imposed by the crisis on the strategic options available to decision makers.

This paper calls attention to the importance of researching radical manufacturing strategies, and briefly identifies the options available to practitioners. Of interest to practitioners, this paper shows that it is possible to identify business distress with the use of predictive models. One of the messages of this paper is that the advantage of using failure prediction models is that they alert manufacturers to problems before a serious crisis develops, and that there are only limited manufacturing strategy options once a crisis strikes.

References

Adam, E.E. and P.M. Swamidass. "Assessing Operations From A Strategic Perspective," Journal of Management, 15 (2), Summer, 1989, 181-203.

Altman, E.I. "Financial Ratios, Discriminant Analysis and the Prediction of Corporate Bankruptcy." The Journal of Finance, September 1968, p. 589-609.

Anderson, J.C., G. Cleveland, and R.G. Schroeder. "Operations Strategy: A Literature Review," Journal of Operations Management, 8 (2), April, 1989, 133-158.

Blum, M. "Failing Company Discriminant Analysis." Journal of Accounting Research, Spring 1974, 1-25.

Deakin, E.B. "A Discriminant Analysis of Predictors of Business Failure." Journal of Accounting Research, Spring 1972, 167-179.

El Hennaway, R.H.A. and R.C. Morris. "The Significance of Base Year in Developing Failure Prediction Models," Journal of Business Finance and Accounting, 1983, 209-223.

Moses, D. and S.S. Liao. "On Developing Models for Failure Prediction," The Journal of Commercial Bank Lending, March 1987, p. 27-38.

Porter, M. Competitive Strategy: Techniques for Analyzing Industries and Competitors, New York: Free Press, 1980.

Swamidass, P.M. "The Role of Inventory and Cost in Manufacturing Business Failures: An Autopsy of Failed Manufacturers," Working Paper, College of Business and Public Administration, University of Missouri, Columbia, MO 65211, 1989.